Somatic Cell
Genetics

NATO ADVANCED STUDY INSTITUTES SERIES

A series of edited volumes comprising multifaceted studies of contemporary scientific issues by some of the best scientific minds in the world, assembled in cooperation with NATO Scientific Affairs Division.

Series A: Life Sciences

Recent Volumes in this Series

This series is published by an international board of publishers in conjunction with NATO Scientific Affairs Division

A Life Sciences	Plenum Publishing Corporation
B Physics	London and New York
C Mathematical and Physical Sciences	D. Reidel Publishing Company Dordrecht, The Netherlands and Hingham, Massachusetts, USA
D Behavioral and Social Sciences	Martinus Nijhoff Publishers The Hague, The Netherlands
E Applied Sciences	

Somatic Cell Genetics

Edited by
C. Thomas Caskey
and
D. Christopher Robbins

Howard Hughes Medical Institute
Baylor College of Medicine
Houston, Texas

SPRINGER SCIENCE+BUSINESS MEDIA, LLC

Library of Congress Cataloging in Publication Data

NATO Advanced Study Institute of Somatic Cell Genetics (1981: Algarve, Portugal)
Somatic cell genetics.

(NATO advanced study institutes series. Series A, Life sciences; v. 50)
Bibliography: p.
Includes index.
1. Cytogenetics—Congresses. I. Caskey, C. Thomas. II. Robbins, D. Christopher. III.
North Atlantic Treaty Organization. Division of Scientific Affairs. IV. Title. V. Series
[DNLM: 1. Cytogenetics—Congresses. 2. Hybrid cells—Congresses. QH 605 N281s 1981]
QH430.N37 1981   599.08′73223   82-7604
                                                                                    AACR2

ISBN 978-1-4684-4258-8   ISBN 978-1-4684-4256-4 (eBook)
DOI 10.1007/978-1-4684-4256-4

Proceedings of a NATO Advanced Study Institute of Somatic Cell
Genetics, held May 31–June 12, 1981, in Algarve, Portugal

© 1982 Springer Science+Business Media New York
Originally published by Plenum Press New York in 1982
Softcover reprint of the hardcover 1st edition 1982

PREFACE

This book represents a selected group of manuscripts from lecturers participating in the NATO/Gulbenkian Foundation sponsored course on Somatic Cell Genetics held May 31 to June 12, 1981 in the Hotel Montechoro in the Algarve of Portugal. The text will provide those students who could not attend the meeting with a current survey of important advances in the field of Somatic Cell Genetics. It is not possible to recapture here all the lectures, seminar discussions, student and faculty interactions, the ambience of the Algarve and the time devoted exclusively to scientific discussion. In summary, I feel that this book is good, but the scientists, the students, and the entire course were better.

Somatic Cell Genetics is a broad subject area and one which has contributed significantly to our understanding of the mammalian cell. Drs. Caskey, Buttin, Siminovitch, and Lechner elected in designing the course to focus on the results obtained with cultured animal cells.

Animal cells can alter their phenotype in culture, and thus we addressed the questions of mutation, chromosome alteration, and differentiation. Significant new advances were reported in the molecular delineation of mutational events at a wide variety of single gene loci (HPRT, AA-tRNA synthetases, protein biosynthesis, etc.) in hamster cells particularly. In some cases cells alter their phenotype by amplification of specific genes. The pioneer work related to methotrexate resistance has led this field. In still other cases such as the liver cell, differentiation events can be reproduced in culture but are not delineated at the molecular level.

The progress of cell hybridization, antibody-producing hybridomas, recombinant DNA cloning, and gene transfer have provided somatic cell geneticists with a new technology for achieving rapid advances. These advances include: establishment of a human gene map; molecular study and cellular expression of cloned genes; identification of unknown genes which have a phenotypic character expressed in a cultured cell (eg. neoplasia); and development of

specialized cells making specific antibodies. Each of these areas
was discussed in the course and the manuscripts here described the
most recent advances in the area.

The Somatic Cell Genetics course was made possible by our oper-
ating committee, faculty, and financial support from NATO and the
Gulbenkian Foundation. It was made operative by the dedicated
efforts of Dr. Maria C. Lechner and Mr. D. Christopher Robbins.
Mr. Robbins is also to be credited for the careful attention to
this text preparation. To these individuals we are grateful.

                                                              C.T. Caskey

January 1982

CONTENTS

PATHWAY MUTANTS IN MAMMALIAN CELLS AND THEIR CONTRIBUTION TO THE ANALYSIS OF DRUG RESISTANCE

Gérard Buttin, Michelle Debatisse and Bruno Robert de Saint Vincent

Institut de Recherche en Biologie Moléculaire (CNRS)
Tour 43 - 2, place Jussieu
75251 PARIS Cedex 05 (France)

INTRODUCTION

The isolation by Beadle and Tatum (1) of Neurospora mutants with altered anabolic activity was a landmark in biology ; the mutant hunt which followed these pioneer experiments became an essential tool in the reconstruction of the main pathways which govern the metabolic activity of both prokaryotic and eukaryotic microorganisms. The analysis of metabolic steps in mammals and man did not rely as deeply on the genetic approach but it was largely guided by the information gained from these studies. Today most mammalian anabolic and catabolic pathways have been reconstructed; their enzymes have been extensively purified and controls exerted on their activity by a variety of regulatory effectors have been demonstrated. Yet an important part of the activity of somatic cell geneticists remains devoted to the isolation and characterization of pathway mutants. Their efforts are directed towards solving three main problems: (a) the nature of heritable variation in cultured mammalian cells, which was a source of considerable speculations, has been clarified in a limited number of systems only; (b) the availability of genetic markers necessary for the selection of cells which have acquired foreign genetic information remains a limiting factor in cell hybridization and gene transfer, and, more generally, in all rapidly expanding experiments of cellular engineering; (c) the physiological importance of an enzyme cannot be inferred from "in vitro" experiments: they do not take in account the network structure of most metabolic pathways and supply at best incomplete information on the influence which a variety of regulatory

1

effectors can exert on the activity of an enzyme. The turn off of
this activity by a specific inhibitor or by a mutation is required,
but most inhibitors have multiple targets, some of which may be
unsuspected.

Since the two first aspects of the utilization of somatic
cell mutants will be considered elsewhere in this volume, this
article will be primarily devoted to illustrate on two examples -
arabinofuranosylcytosine (araCyt) resistance and coformycin
resistance of Chinese hamster fibroblasts - the unique advantage
of pathway mutants for physiological analysis. The pathways under
investigation belong to those which generate nucleic acid
precursors. These are very complex systems since both "de novo"
biosynthetic pathways and the so-called "salvage" pathways for
utilization of preformed bases and nucleosides can contribute to
the replenishment of the nucleoside-triphosphates pools."Suicide"
selection techniques have been exploited with great succes to
isolate mutants lacking enzymes of the "de novo" biosynthetic
pathways (2). Toxic nucleoside analogs are powerful tools for the
direct and simple recovery of mutants bearing defects in the
activity of "salvage" enzymes phosphorylating preformed bases and
nucleosides. But they can also provide insights into the function
and into the physiological regulation of a variety of enzymes
involved in nucleotide metabolism, including those belonging to
interconversion pathways. Moreover, as will be discussed, genetic
analysis may contribute to a better understanding of the mechanisms which permit dividing cells to escape destruction by toxic
analogs of common use in chemotherapy.

MUTATIONS ALTERING PYRIMIDINE METABOLISM: ARACYT RESISTANCE

Dose-dependent Pattern of Drug Resistance

When mutagenized Chinese hamster cells of the CCL39 line
were plated in medium containing 5µg/ml of araCyt, all colony-
forming cells tested for their resistance level were found able
to form colonies at concentrations as high as 50µg/ml ("high
resistance" phenotype). When the selective medium contained no
more than 0.5µg/ml of araCyt, such mutants were indeed recovered,
but a new class of resistant clones was observed : they are
unable to form colonies when araCyt concentration reaches 1µg/ml
("low resistance"phenotype) (3).

Loss of dCyd-araCyt-Kinase Activity in "High Resistance" Mutants

All "high resistance" mutants examined were found deficient
in araCyt-kinase activity (3). The same biochemical defect has
been characterized in resistant clones selected in culture from
different mammalian lines, and genetic and biochemical analysis

agree to identify the altered enzyme as a deoxycytidine-kinase.
(designated below as "dCyd-AraCyt-kinase") (4,8). The observation
that an important level of dCyd-kinase was preserved in araCyt-
kinase deficient Chinese hamster fibroblasts, disclosed the
presence in these cells of an active mitochondrial isozyme of
dCyd-kinase, which does not utilize araCyt as a substrate (8).

Expansion of the dCTP Pool in "Low-Resistance" Mutants

The class of "low resistance" mutants is of greater complexity but they all (3,9) appear to share two additionnal phenotypic
properties : their growth is not inhibited by an excess of
thymidine (dThyd) and they have an expanded pool of dCTP. Mutants
which exhibit the same pattern of low araCyt resistance and dCTP
pool expansion can be isolated as resistant to the "dThd block"
(3). Actually, the enlargment of the dCTP pool is expected to
generate the resistance to both nucleosides (fig. 1): (a) dCTP
act both as a competitor of araCTP at the level of nucleic acid
polymerases and as a feed back inhibitor (10) of dCyd-araCyt-
kinase, the first enzyme which converts the toxic nucleoside to
its active form. Expansion of the dCTP pool is therefore expected
to decrease the utilization of exogenous araCyt and to reduce
incorporation of exogenous dCyd and this is indeed observed in
the mutants.

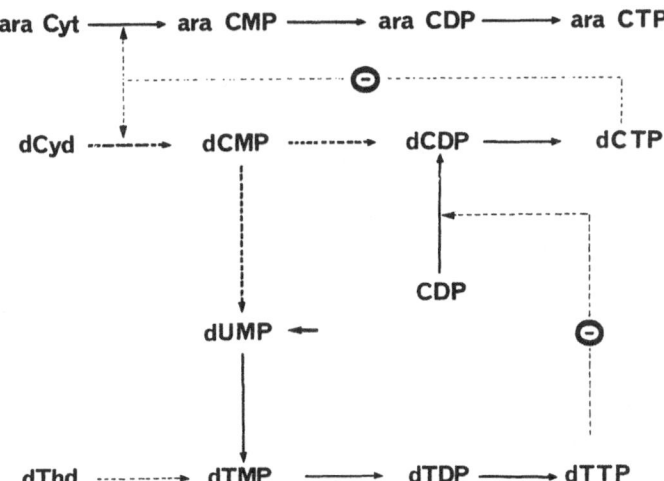

Fig.1. Pathways of pyrimidine deoxyribonucleotide and
araCyt nucleotide synthesis in CCL39.
The ⊖ signs designate negative feedback controls.

(b) the dThd block is well known to be the consequence of dCTP depletion arising when excess dTTP inhibits CDP reductase; it is released by exogenous supply of dCyd and may be avoided by mutations leading to an abnormally high supply of endogenous dCTP.

Enzyme Alterations Leading to dCTP Pool Expansion

CDP-reductase alterations. The highly regulated ribonucleotide-reductase is an obvious possible target for mutations altering the rate of dCTP production. The properties of a mutant isolated from mouse fibroblasts through a multistep selection procedure are consistent with this expectation: this line which is both highly resistant to araCyt and resistant to dThd exhibits an increased level of CDP-reductase, a decreased sensitivity of this enzyme to dATP and an araCyt-kinase deficiency, but the contribution of these multiple alterations to the resistance pattern has not been clarified (11). Recently, mutants selected for their resistance to aphidicolin have been shown to be jointly resistant to araCyt and dThd and characterized as bearing a CDP-reductase desensitized to the inhibitory action of dATP (12).

CTP-synthetase alterations. When we examined however the properties of CDP-reductase in extracts from a set of "low resistance" mutants, we were unable to detect any alteration: the specific activity of the enzyme was identical in wild-type cells and in the mutants, as was its sensitivity to both dATP and dTTP (3).

Another clue came from the observation (Table 1) that in some "low resistance" mutants isolated as resistant to 0.5μg/ml

Table 1. Incorporation of cytidine and deoxycytidine by araCyt-resistant mutants of CCL39. (Incorporation of ^{3}H-labelled nucleosides into TCA precipitable material, expressed as percentage of the incorporation by the wild-type line)

Cell line	dCyd incorporation	Cyd incorporation	AraCyt dCyd kinase activity
CCL39(w.t.)	100	100	+
1A1	21	90	−
1A2	2	130	+
1A3	5	23	+
T1	7	19	+

of araCyt (1A3) or as resistant to 1mM dThd (T1), not only exogenous dCyd but also exogenous cytidine (Cyd) incorporation into
macromolecules was severely reduced. This was not the case for
other mutants such as 1A1, which exhibit the same pattern of
resistance to both nucleosides, nor for a mutant (1A2) selected
for resistance to 0.5µg/ml of araCyt but actually resistant to
50µg/ml of the drug and identified as an araCyt-dCyd-kinase
deficient clone. The previous observations that dCTP pool expansion imposed a reduced incorporation of exogenous dCyd suggested
that reduced Cyd incorporation in 1A3 and T1 might similarly be
the consequence of an abnormally high pool of CTP. This hypothesis
was supported by direct ribonucleoside-triphosphate pool measurements (Table 2-A), which also revealed that in 1A3 and T1, there
is no increase of the intracellular concentration of UTP, the
immediate precursor of CTP (13). These measurements strongly
suggested that the primary biochemical defect in these lines was
an altered regulation of CTP-synthetase activity or synthesis.
In wild-type or mutant cell extracts, the level of CTP synthetase
is actually identical; but the well known inhibition exerted on
the activity of the enzyme by its product is considerably lower
in T1 than in its CCL39 parent (fig.2): a 0.2 mM concentration
of CTP completely inhibits the partially purified wild-type
enzyme, but about 50% of the activity is preserved in the T1
mutant under these conditions. The desensitization of CTP-synthetase can indeed account for the expansion of the CTP pool and
presumably of the dCTP pool derived from it. Therefore, it offers

Table 2. Triphosphate pool levels and araCyt resistance of wild-
         type, resistant, and revertant cell lines

Cell line	Triphosphate pool (nmol/10^6 cells)		AraCyt (µg/ml) resistance (LD90)
	CTP	dCTP	
A.			
CCL39 (w.t.)	1.1	0.36	0.05
1A1	1.0	1.15	0.8
1A2	1.0	0.25	50
1A3	3.4	1.50	0.7
T1	3.7	1.60	0.7
B.			
TA43	3.5	1.50	0.8
TA43 rev1	1.0	0.45	0.08
TA43 rev2	1.2	0.55	0.10
TA43 rev3	1.0	0.40	0.08

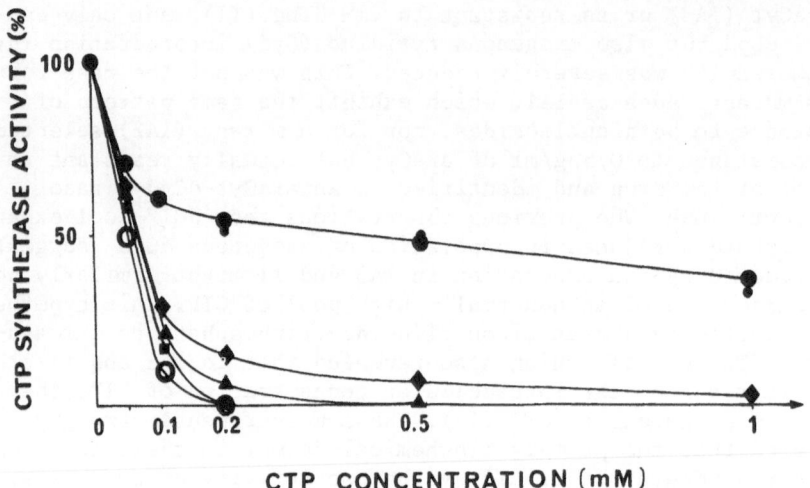

Fig.2.   Inhibition of CTP-synthetase activity by CTP in wild-
         type, mutant and revertant lines; O: CCL39; ●: T1;
         ●: TA43; ▲: TA43 rev$_a$; ◆: TA43 rev$_b$; ■: TA43
         rev$_c$

a satisfactory explanation to the low resistance phenotype of the
T1 mutant. It should be stressed however that one has to be
cautious in drawing such a conclusion as long as it has not been
unequivocally shown that reversion of the identified biochemical
defect is sufficient to restore the wild-type resistance level.
This word of caution is not gratuitous, as illustrated in this
system by the properties of the 1A1 mutant line (3,13,14). In
this low resistance mutant with a normal CTP pool we noticed that
dCMP-deaminase is inactive. On a simple logical basis, this defect
might account for the joint resistance to araCyt and dThd, since
dCMP-deaminase activity diverts a potential source of dCTP towards
the dTTP pool. But the same enzymatic deficiency was identified
in a variety of sublines isolated from CCL39 on the basis of
selections with no obvious relationship to dCMP metabolism. These
independent lines preserve wild-type sensitivity to araCyt and
dThd, indicating that dCMP deaminase deficiency is not responsible
or not sufficient "per se" to determine the resistant phenotype.

    A definitive proof that CTP-synthetase alteration is respon-
sible for the resistant phenotype of T1 was brought by the
isolation of revertants which simultaneously regained wild-type

sensitivity of the enzyme to CTP, and wild-type sensitivity to
both araCyt and dThd (13).

This approach took advantage of two observations: (a) a
clone (TA43) deficient in dCMP-deaminase was isolated by chance
from the T1 line. The mechanism which generates at a surprisingly
high frequency dCMP-deaminase-deficient variants from the CCL39
line is not known but its genetic or epigenetic character is
irrelevant to this discussion. (b) In dCMP-deaminase deficient
lines, expansion of CTP pool, which can be brought about either
by a genetic alteration or by growing the cells in medium supplemented with CTP, inhibits the UDP-reductase pathway (14). This
property makes dCMP-deaminase deficiency a conditionnally lethal
trait, when coupled to a mutation expanding the CTP pool: in
such double mutants, the two endogenous pathways which generate
dUMP are shut off, as shown in fig.3, and cells rely on exogenous
dThd or dUrd supply for dTTP synthesis and for their division.
In medium devoid of these nucleosides, the double mutant dies,
but surviving clones can be isolated; they are expected to
include revertants which have recovered either an active dCMP-
deaminase or a reduced CTP pool. This expectation was fullfilled:
3 clones, which do no longer manifest dThd auxotrophy were isolated from TA43; they have a wild-type CTP pool level: all three
simultaneously recovered essentially wild-type sensitivity of
CTP synthetase to CTP (fig.2), wild-type level of the dCTP pool
and wild-type sensitivity to araCyt (Table 2) and to dThd (13).

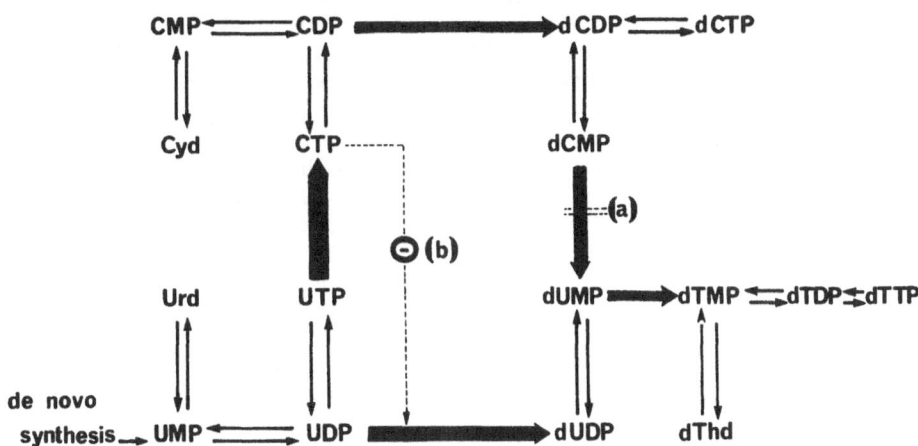

Fig.3. Pathways of pyrimidine nucleotide synthesis in CCL39.
        (a): block imposed by dCMP-deaminase deficiency;
        (b): block imposed by CTP overproduction.

The enzyme alteration responsible for dCTP pool expansion in other low resistance mutants such as 1A1 remains to be discovered. But a remarkable property of all "low resistance" lines which we examined is the semi-dominant expression of resistance in hybrids generated by fusion with sensitive cells. This is in striking contrast to the expression of araCyt resistance caused by araCyt dCyd-kinase deficiency, which is recessive in hybrids (3). Some possible implications of this property will be discussed below.

MUTATIONS ALTERING PURINE METABOLISM: COFORMYCIN RESISTANCE

A markedly different mechanism of resistance to a nucleoside analog was disclosed during the course of investigations on the regulation of cell proliferation by preformed purines and their analogs (15;16).

Adenine is not toxic to Chinese hamster fibroblasts (line GMA32, an araCyt-dCyd-kinase derivative of CCL39) growing in regular medium, unless it is added at 1mM concentrations or above. Coformycin, an adenosine analog known to inhibit adenosine-deaminase at concentrations as low as 0.02µg/ml, is not toxic either even if added at 2µg/ml. However, if adenine at 10^{-5} M is added to medium containing 0.5µg/ml of coformycin, the cells die. The mixture (HC + A) of coformycin at this high concentration (HC) and adenine (A) is not toxic to APRT⁻ cells, indicating that adenine must be phosphorylated to AMP in order to contribute to lethality. Besides, the wild-type cells are rescued when the HC + A medium is supplemented with hypoxanthine, while HGPRT⁻ cells are not: this suggests that, in HC + A medium, the cells die of IMP or GMP starvation, since they can synthesize adenylic nucleotides from adenine. This conclusion is supported by the observation that the very same concentration of adenine kills the cells and feed back inhibits the endogenous purine biosynthetic pathway. The simple model suggested by these results is that, when coformycin is present, the production of IMP from adenylic purines is decreased to a level unsufficient to sustain cell growth. Since two enzymes - adenosine-deaminase and AMP-deaminase - can divert adenylic purines to generate IMP, this further suggested that the drug - as such or via metabolic derivatives - inhibits both enzymes (fig. 4).

To check this hypothesis, mutants able to grow in HC + A medium were sought for. These clones distribute into two classes (16):

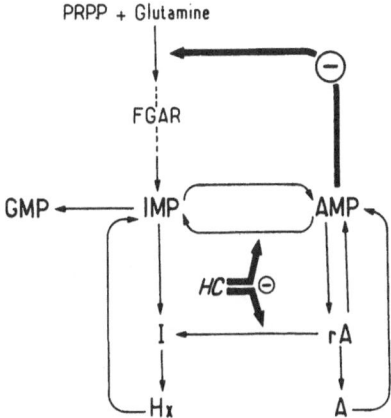

Fig.4. Pathways of purine nucleotide synthesis in
        hamster fibroblasts. The ⊖ signs designate
        negative controls exerted by AMP, or coformycin
        at 0.5μg/ml (HC).

Mutants with Altered Regulation of "de novo" Purine Biosynthesis

     Class 1 mutants are characterized by their property to be
unable to grow in the selective medium if azaserine - an inhibitor
of "de novo" IMP synthesis - is further added. This indicates
that the proliferation of these cells in the selective medium is
possible because they somehow escape the feedback inhibition of
the endogenous pathway. To check this point we examined the
influence of adenine on the accumulation of N-formylglycine
amide ribotide (FGAR) - which measures (17) the activity of the
endogenous pathway of purine synthesis - in the parental line
and in some of these mutants. As expected, 10^{-5}M adenine completely blocks purine synthesis in the parental cells but not in
the class 1 mutants. The mechanisms for the release of inhibition
are not known, but it is interesting that in one mutant, half
inhibition is maintained. This is the situation predicted if the
mutation alters one autosomal gene coding for the first enzyme
of the "de novo" pathway, which - in wild-type cells - is feed-
back inhibited by AMP.

Mutants with Increased Activity of a Purine Interconversion Enzyme

In contrast to class 1 mutants, class 2 mutants grow in the selective HC + A medium independently of the activity of the endogenous pathway: selection in this medium supplemented with azaserine can be utilized to recover only mutants of this class. All class 2 mutants examined (3/3) had normal adenosine-deaminase activity but 7-13 fold the wild-type level of AMP-deaminase activity, suggesting that resistance is the manifestation of an increased AMP-deaminase activity, compensating for partial inhibition of this enzyme by coformycin. In agreement with this interpretation, we observed that coformycin slowly inhibits "in vivo" AMP-deaminase of both wild-type and mutant cells. Inhibition reaches a plateau, leaving in the mutant an AMP-deaminase activity comparable to the activity found in wild-type cells growing in the absence of the drug. Interestingly, the level of resistance

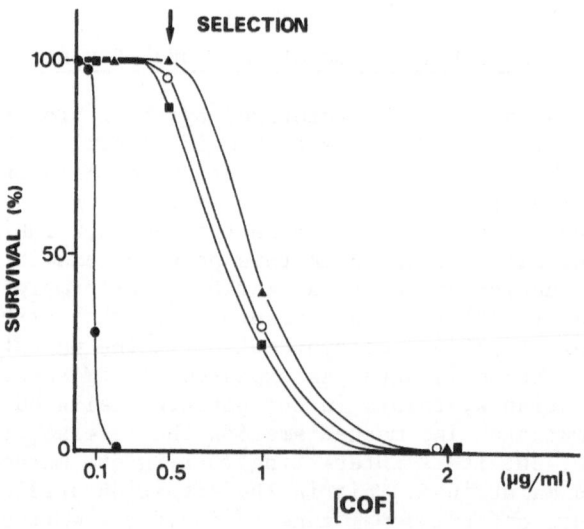

Fig.5.   Plating efficiency of mutants, selected for their
           resistance to coformycin (0.5μg/ml) + adenine, in
           the same medium supplemented with increasing
           concentrations of the analog.

of the mutants to coformycin - when measured in several clones soon after their isolation - was found to be precisely the drug concentration used for their selection (fig. 5). Resistance in these clones exhibits some other remarkable properties:

- resistance can be progressively increased by successive steps of selection in the presence of increasing concentrations of the drug. AMP-deaminase activity increases in parallel to resistance levels, reaching 150 fold the wild-type level in a clone recovered from a final selection in medium containing 25µg/ml of coformycin (and 10^{-5}M adenine) (M. Debatisse,unpublished).

- resistance is unstable (15). Prolonged growth in regular medium yields cells which have recovered both full sensitivity to HC + A medium and wild-type AMP-deaminase activity. From one unstable clone, a stable subclone with an intermediate level of resistance was recovered; it has an intermediate level of AMP-deaminase activity (Table 3).

Table 3. Resistance level and AMP-deaminase activity of the unstable mutant HC_{10}-61 (selected in 5µg/ml of coformycin) and of its stabilized subclone HC_{10}-61-S_1, after various periods of growth in regular medium

Cell line	Transfer number in regular medium before plating	Plating efficiency in medium (A + azaserine) + coformycin (µg/ml)			AMPD specific activity
		0	0.5	5	
GMA32 (w.t.)	–	100	0	0	405
HC_{10}-61	0	100	100	100	15690
	7	100	100	60	–
	29	100	25	0	620
HC_{10}-61-S_1	5	100	87	0	4950
	36	100	93	0	5170
	44	100	88	0	5156

    - resistance is a dominant trait in cell hybrids (M.Debatisse,
unpublished)

    - resistance is associated with overproduction or stabiliza-
tion of AMP-deaminase protein. As illustrated in fig. 6, the
protein band characteristic of the enzyme is observable in the
SDS-PAGE electrophoresis pattern of an hyperactive mutant extract,
not of a wild-type extract. Moreover, this electrophoretic

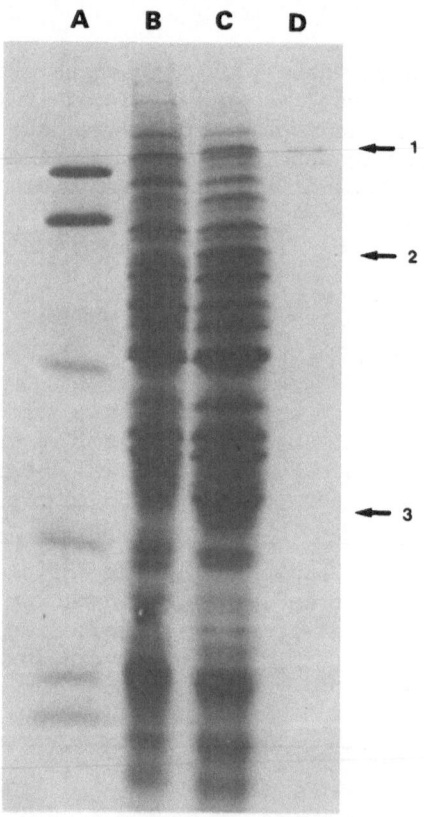

Fig. 6.   SDS-PAGE electrophoretic pattern of wild-type and
           mutant cell extracts. Lanes: A: Mol weight markers
           (from top to bottom: phosphorylase B: 93.000;
           bovine albumin: 66.000; pepsin: 34.700; pepsino-
           gen: 24.000; β-lactoglobulin: 18.400; lysozyme:
           14.300 -   B: CCL39 extract -   C: HC_{50}-611 extract
           (resistant to 25μg/ml coformycin) - D: AMP-deami-
           nase purified from Chinese hamster cells. Arrows:
           protein bands "amplified" in the mutant.

pattern reveals the presence in the mutant extract of at least two additional protein bands which are not detected in the parental cell extract and may represent additional unidentified proteins overproduced by this mutant.

The properties of the class 2 resistant mutants are strongly reminiscent of those reported for unstable methotrexate resistant mutants, which were subsequently shown to bear multiple copies of the gene coding for dihydrofolate-reductase (DHFR), the target enzyme for this tight binding inhibitor (18). The similarity between the two systems is reinforced by the observation of a "homogeneously staining region" (HSR) on one abnormally long chromosome present in the cells of a clone exhibiting 40 fold the wild-type AMP-deaminase activity and in its subclones. The HSR is located at the point at which a small chromosome is translocated to one long chromosome of the pair number 1. In the methotrexate system HSRs have been observed on chromosomes of subclones with stable resistance and identified as the site of intrachromosomal DHFR amplified genes (19). Work in progress should establish whether, as seems to be a reasonable guess, resistance to the toxic mixture of adenine and coformycin is the manifestation of amplification of the AMP-deaminase gene.

CONCLUDING REMARKS

We summarized here the results of attempts to identifying mutants isolated on the basis of their sensitivity to toxic nucleoside analogs. The strategy developed to reach this goal relied on the known biochemistry of enzymes contributing to the pathways of nucleic acid precursor metabolism : the analysis of genetic defects altering other biochemical pathways requires, in each particular case, a specific approach. But some general remarks can be drawn from this and similar studies. One is the diversity of biochemical mechanisms which the genetic approach can disclose as being at the origin of a simple phenotypic alteration, such as resistance to a drug. As exemplified above, they include the turn off of an enzyme activity, but also the alteration of a regulatory site on an enzyme molecule, and the overproduction(or stabilization) of a target enzyme. It is also clear that modest changes in the concentration of a toxic agent during the selection procedures can markedly modify the pattern of resistant mutants recovered, and yield predominantly dominant or recessive clones.

These observations may be of interest when one considers the problem of drug resistance arising during prolonged chemo-therapy in man. The appearance of resistant clones from tumor cells - which are at least diploid and frequently of an even

higher ploidy level - is expected to be usually the consequence of
a dominant change. Among dominant changes, special attention
should be paid, we believe, to those leading to the slow emergence
of clones with increasing resistance from cultures repeatedly
challenged with a drug at concentrations slightly above the dose
lethal to the average cell population. The mere existence of such
mutants suggests that one should perhaps prefer, when it is
possible, the utilization of high concentrations of the analog
during short periods to its repeated administration at lower, yet
toxic concentrations.

The information gained from experiments carried out on somatic
cell lines cultured "in vivo" must indeed be considered with some
caution : as an example, a nucleoside analog, such as coformycin,
exhibits no cytotoxic properties when added to regular cell growth
medium but kills dividing fibroblasts when the medium contains a
very low concentration of a natural metabolite - adenine - which
may be present in biological fluids. Thus, although coformycin
would be classified as non-toxic - even at high concentrations -
in some standardized assays for antimetabolites carried out "in
vitro", it may exhibit marked cytotoxic activity to the dividing
cells of all or some tissues within the organism. One should also
keep in mind that cells of different histotypic origin or isolated
from different animal species may differ both qualitatively and
quantitatively in their enzymatic equipment. Therefore, the
pattern of drug resistant mutants recovered from a "model" cell
system may be specific to this system : the resistance to AraCyt
of human tumor cells with a very active AraCyt-dCyd deaminase
activity (20,21) may usually be caused by alterations different
from those observed in rodent fibroblasts which lack this activity.
But somatic cell genetics remains a unique tool to disclose the
existence and to demonstrate the physiological importance of
regulations of potential interest for the development of new
therapeutic strategies. This can be illustrated by the above
mentioned control which CTP exerts on the activity of the UDP-
reductase pathway, thereby making CTP overproducing fibroblasts
dependent on the activity of dCMP-deaminase for their DNA
synthesis. CTP overproduction has been reported to be a property
of rapidly growing hepatomas and kidney tumor cells (22). It will
be of special interest to determine whether these cells also rely
on dCMP-deaminase activity for their dUMP production. If this
were the case, then two enzymes - dCMP-deaminase and dTMP-
synthetase, acting sequentially for DNA synthesis - would
constitute independent targets for inhibitors: simultaneous
administration of inhibitors of the two enzymes would be a most
efficient way to avoid the multiplication of mutant cells which
would escape therapy by either drug. In a more general way,
inhibitors of dCMP-deaminase might constitute an important class
of antitumor drugs.

ACKNOWLEDGEMENTS

The work summarized here was supported in part by the Ligue Nationale Française contre le Cancer, the Fondation pour la Recherche Médicale Française, the DGRST (Grant n°79.7.1036), and the Commissariat à l'Energie Atomique. A generous gift of coformycin by Drs H. Umezawa and I. Kitasoto is gratefully acknowledged.

REFERENCES

1. G.W. Beadle & E.L. Tatum, Genetic Control of Biochemical Reactions in Neurospora, Proc. Natl. Acad. Sci. USA **27**:499 (1941).

2. M. Irwin, D.C. Oates & D. Patterson, Biochemical Genetics of Chinese Hamster Cell Mutants with Deviant Purine Metabolism: Isolation and Characterization of a Mutant Deficient in the Activity of Phosphoribosylaminoimidazole Synthetase, Somat. Cell Genet. **2**:203 (1979).

3. B. Robert de Saint Vincent & G. Buttin, Studies on 1-β-D-arabinofuranosyl cytosine resistant mutants of Chinese hamster fibroblasts. III. Joint resistance to arabinofuranosyl cytosine and to excess thymidine – a semi dominant manifestation of deoxycytidine triphosphate pool expansion, Somat. Cell Genet. **5**:67 (1979).

4. M.Y. Chu & G.A. Fischer, Comparative studies of leukemic cells sensitive and resistant to cytosine arabinoside, Biochem. Pharmacol. **14**:333 (1965).

5. A.W. Schrecker, Metabolism of 1-β-D-arabinofuranosylcytosine in leukemia L 1210 : nucleoside and nucleotide kinases in cell-free extracts, Cancer Res. **30**:632 (1970).

6. D. Drakowsky & W. Kreis, Studies on drug resistance II. Kinase patterns in P815 neoplasms sensitive and resistant to 1-β-D-arabinofuranosylcytosine, Biochem. Pharmacol. **19**:940 (1970).

7. W. Kreis, D. Drakowsky & H. Borberg, Characterization of protein and DNA in P815 cells sensitive and resistant to 1-β-D-arabinofuranosylcytosine, Cancer Res. **32**:696 (1972).

8. B. Robert de Saint Vincent & G. Buttin, Studies on 1-β-D-arabinofuranosyl-cytosine-resistant mutants of Chinese hamster fibroblasts. A mitochondrial deoxycytidine kinase devoid of activity on arabinocytosine, Eur. J. Biochem. **37**:481 (1973).

9. R.L. Monparler, M.Y. Chu & G.A. Fischer, Studies on a new mechanism of resistance of L 5178Y murine leukemia cells to cytosine arabinoside, Biochim. Biophys. Acta **161**:481 (1968).

10. P.G.W. Plagemann, R. Marz & R.M. Wohlhueter, Transport and
     Metabolism of deoxycytidine and 1-β-D-arabinofuranosylcytosine
     into cultured Novikoff rat hepatoma cells, relationship to
     phosphorylation, and regulation of triphosphate synthesis,
     Cancer Res. **38**:978 (1978).
11. M. Meuth & H. Green, Alterations leading to increased
     ribonucleotide reductase in cells selected for resistance to
     deoxynucleosides, Cell **3**:367 (1974).
12. D. Ayusawa, K. Iwata & T. Seno, Alteration of ribonucleotide
     reductase in aphidicolin-resistant mutants of mouse FM3A cells
     with associated resistance to arabinosyladenine and
     arabinosylcytosine, Somat. Cell Genet. **7**:27 (1981).
13. B. Robert de Saint Vincent & G. Buttin, Studies on 1-β-D-
     arabinofuranosyl cytosine resistant mutants of Chinese hamster
     fibroblasts. IV. Altered regulation of CTP-synthetase
     generates arabinosylcytosine and thymidine resistance,
     Biochim. Biophys. Acta **610**:352 (1980).
14. B. Robert de Saint Vincent & G. Buttin, The modulation of the
     thymidine triphosphate pool of Chinese hamster cells by dCMP
     deaminase and UDP reductase : thymidine auxotrophy induced by
     CTP in dCMP deaminase deficient lines, J. Biol. Chem. **255**:
     162 (1980).
15. M. Debatisse & G. Buttin, The control of cell proliferation
     by preformed purines : a genetic study. II. Pleiotropic
     manifestations and mechanisms of a control exerted by adenylic
     purines on PRPP synthesis, Somat. Cell Genet. **3**:513 (1977).
16. M. Debatisse, M. Berry & G. Buttin, The potentiation of
     adenine toxicity to Chinese hamster cells by coformycin :
     suppression in mutants with altered regulation of purine
     biosynthesis or increased adenylate-deaminase activity, J.
     Cell Physiol. **106**:1 (1981).
17. D. Patterson, F.T. Kao & T.T. Puck, Genetics of somatic
     mammalian cells : biochemical genetics of Chinese hamster
     mutants with deviant purine metabolism, Proc. Natl. Acad.
     Sci. USA **71**:2057 (1974).
18. R.T. Schimke, R.J. Kaufman, F.W. Alt & R.F. Kellems, Gene
     amplification and drug resistance in cultured murine cells,
     Science **202**:105 (1978).
19. J.H. Nunberg, R.J. Kaufman, R.T. Schimke, G. Urlaub & L.A.
     Chasin, Amplified dihydrofolate reductase genes are localized
     to a homogeneously staining region of a single chromosome in
     a methotrexate-resistant Chinese hamster ovary cell line,
     Proc. Natl. Acad. Sci. USA **75**:5553 (1978).
20. D.H.W. Ho, Distribution of kinase and deaminase of 1-β-D-
     arabinofuranosylcytosine in tissues of man and mouse, Cancer
     Res. **33**:2816 (1973).

21. T.C. Chou, Z. Arlin, B.D. Clarkson & F.S. Philips, Metabolism of 1-β-D-arabinofuranosylcytosine in human leukemic cells, <u>Cancer Res.</u> **37**:3561 (1977).

22. J.C. Williams, H. Kizaki, G. Weber & H.P. Morris, Increased CTP synthetase activity in cancer cells, <u>Nature</u> **271**:71 (1978).

ISOLATION AND ANALYSIS OF CHINESE HAMSTER CELLS CARRYING FORWARD
AND REVERSE MUTATIONS IN THE HYPOXANTHINE-GUANINE
PHOSPHORIBOSYLTRANSFERASE LOCUS

Raymond G. Fenwick, Jr., David S. Konecki
and C. Thomas Caskey

Howard Hughes Medical Institute Laboratories, Depart-
ments of Medicine, Cell Biology and Biochemistry
Baylor College of Medicine, Houston, Texas 77030   USA

INTRODUCTION

In studies of somatic cell genetics, mutagenesis, cell fusion,
etc., the locus for hypoxanthine-guanine phosphoribosyltransferase
(HGPRT) has received considerable attention (see Caskey and Kruh
for a review[1]). The product of this gene in various mammalian
systems is a protein of about 25,000 molecular weight which associates into oligomers that catalyze the formation of IMP or GMP from
5-phosphoribosyl pyrophosphate (PP-Ribose-P) plus hypoxanthine or
guanine. A number of purine analogs such as 8-azaguanine and
6-thioguanine also serve as substrates for the enzyme and are converted to toxic nucleotides which kill normal cells and enable the
isolation of HGPRT-deficient clones of cultured cells[2]. Although the enzyme is normally not required for cell growth, agents
such as aminopterin can be used to block de novo synthesis of purine nucleotides thus making cellular multiplication dependent on
the presence of hypoxanthine and HGPRT activity and providing a
reverse selective system for the locus[3]. The realization that
HGPRT-deficiency is the biochemical defect in human patients with
the X-linked Lesch-Nyhan Syndrome[4] localized the locus for the
enzyme to the X chromosome and focused additional attention on the
gene and its product.

Even though forward and reverse mutations affecting the HGPRT
locus of mammalian cells have been studied for the past 20 years,
analysis of those mutational events has been limited because it has
not been possible to conduct recombination studies in cultured
somatic cells. However, the advent of recombinant DNA technologies
has opened new approaches for the precise characterization of genetic alterations which should enable us to localize mutations in

19

the HGPRT gene. In this presentation, we will review the efforts
of our laboratory to study forward and reverse mutation in the
HGPRT locus and describe recent events which will make possible
analysis of those alterations at the level of the DNA sequence.

FORWARD MUTATION AT THE HGPRT LOCUS

     Our labortory chose the V79 clone of Chinese hamster lung
fibroblasts for the mutational analysis of the HGPRT locus because
it had been derived from a male fetus[5] and thus started its
history as a cultured line carrying a single copy of the gene for
the enzyme. The initial studies, as reported by Gillin et al.[6]
sought to define the spectrum of subclones obtained when selections
were carried out for resistance to 30μg/ml 8-azaguanine with or
without prior treatment of the cells with mutagens. After two days
of expression time in non-selective medium, the frequency of colony
forming units resistant to the presence of the analog was found to
be 10^{-5} and this was increased 70-fold after mutagenesis with
either N-methyl-N'-nitro-N-nitrosoguanidine (MNNG) or ethyl
methanesulfonate (EMS). In subsequent studies, UV-light[7] and
ICR191E[8] were found to increase the appearance of drug-resistant colonies 20- and 8-fold, respectively. A substantial number
of clones have been isolated from such selective experiments and
subjected to a variety of cellular and in vitro analyses of their
phenotypes.

     From the outset, it has been clear that our collection con-
tained a heterogeneous set of isolates. For instance, we found
that 17 of the first 35 clones to be characterized were able to
grow in counter-selective medium containing hypoxanthine, aminopterin, and thymidine (HAT) which normally prevents growth of HGPRT-
deficient cells[3]. Furthermore, only 11 of those 17 were able
to grow in the presence of a second purine analog, 6-thioguanine,
used to select for HGPRT deficiency.

     Of the various assays we have used to determine the HGPRT phe-
notypes of our clones, one of the most sensitive and informative
has been the incorporation of ^{14}C-hypoxanthine by intact cells in
the presence or absence of aminopterin. As shown in Table 1, the
inhibition of de novo purine synthesis by aminopterin has little or
no effect on hypoxanthine incorporation by our wild-type clone,
RJKO. It does cause a slight decrease in ^{3}H-uridine incorporation. The results of studies with clones resistant to 8-azaguanine
have fallen into two patterns. Clones such as RJK30 and RJK45 do
not incorporate hypoxanthine. Furthermore, their incorporation of
uridine is severely inhibited in the presence of aminopterin. That
is probably a reflection of a starvation for purine nucleotides
caused by the combination of aminopterin and HGPRT deficiency.
Clones such as RJK49 and RJK62 also do not incorporate hypoxanthine
but can be induced to do so by adding aminopterin to the medium.

Table 1. Cellular Assays of HGPRT Activity [a]

Clone	Aminopterin	Acid-insoluble Radioactivity		
		^{3}H-Urd	^{14}C-Hyp	$\frac{Hyp}{Urd}$
RJK0	−	179,652	204,294	1.14
	+	148,824	208,358	1.40
RJK30	−	244,041	1,070	.004
	+	36,824	285	.008
RJK45	−	198,031	833	.004
	+	5,864	78	.014
RJK49	−	334,283	945	.003
	+	150,490	102,910	.68
RJK62	−	338,573	−1,031	−.003
	+	66,741	41,991	.63

[a] Cells were plated in non-selective medium at 10^5 per 60mm
dish and 48 hrs later incorporation of radioactive precursors was
measured during a 5 hr incubation in the presence or absence of
10μM aminopterin as previously described[9].

In the presence of that drug, however, their incorporation of uridine is decreased significantly and the ratio of hypoxanthine to
uridine incorporated by those cells never reaches the values at-
tained by wild-type cells. Thus, the latter isolates have detectable HGPRT activity but it is only functional in the cell when de
novo purine synthesis is inhibited by aminopterin or other
agents[10] and even then it does not seem able to supply enough
purine nucleotides to maintain a normal rate of nucleic acid synthesis as measured by uridine incorporation. We have found that
clones having only a few percent of normal HGPRT activity in cell
free extracts often incorporate more than 50% as much hypoxanthine
as wild-type cells when aminopterin is present.

When the incorporation assay for HGPRT activity was used to
evaluate the enzyme status of our collection of clones resistant to
8-azaguanine, we found that about one-fourth had detectable HGPRT
activity. However, as illustrated in Table 2, the use of mutagens
during isolation of the clones had a profound effect on the frequency of resistant clones that were still able to incorporate
hypoxanthine.A substantial number of isolates which arose after
treatment with the point mutagens MNNG and EMS would incorporate
hypoxanthine but none of the ten spontaneous isolates had that

Table 2. Phenotypic Properties of Clones
Resistant to 8-azaguanine

Inducing Mutagen	Number of Isolates			
	Total [a]	HGPRT$^+$ [b]	HAT^r [c]	CRM$^+$ [d]
None	10	0	–	0/10
MNNG	23	14	12	2/9
EMS	18	5	5	2/13
UV	10	1	NT	0/8
ICR191E	18	1	1	1/17

[a] Presumptive HGPRT mutants were isolated as 8-azaguanine resistant clones with or without prior treatment of the initial population with a mutagen[6].

[b] Cellular enzyme activity measured by incorporation as in
Table 1.

[c] The number of clones which are HGPRT-positive and have
unreduced plating efficiencies in HAT medium.

[d] Fraction of the HGPRT-negative clones tested in which enzyme
protein could be detected by immunological assays.

phenotype. It was also found to be rare among the clones isolated
after treatment with the frameshift mutagen ICR191E or irradiation
with ultraviolet light. Thus, use of mutagens known to cause base
substitutions in lower systems resulted in the isolations of clones
with residual HGPRT activity. This would be consistent with the
introduction of missense mutations by those mutagens and also imply
that the spontaneous or UV-induced isolates might carry more severe
genetic alterations such as deletions or rearrangements. The data
in Table 2 also show that 18 of the 20 isolates which were able to
incorporate hypoxanthine were also able to form colonies in HAT
medium. Thus, it was obvious early in our investigations that many
of our 8-azaguanine isolates retained significant amounts of HGPRT
activity.

The biochemical basis for resistance to both 8-azaguanine and
HAT as well as the aminopterin stimulation of hypoxanthine incorpo-
ration was revealed when we examined the kinetic parameters of
HGPRT from such isolates. The data in Table 3 show that higher

Table 3. Kinetic Properties of HGPRT
from Enzyme Positive Mutants[a]

Clone	Vmax, $\frac{nmol}{mg \cdot min}$	Km or $[S_{0.5}]$, μM	
		Hyp	PP-Ribose-P
RJK0	1.44	1.9	30
RJK3	.32	0.47	50
RJK44	2.44	2.3	360
RJK47	1.43	2.5	230

[a] Enzyme assays were conducted in extracts of wild-type, RJK0, and mutant cells as described previously[9].

concentrations of PP-ribose-P are required to attain maximum rates of HGPRT activity in extracts of the MNNG-induced strain RJK3 or the EMS-induced isolates RJK44 and RJK47 than in extracts of wild-type cells, RJK0. Furthermore, we found that HGPRT from RJK3, 44 and 47, unlike that from wild-type was activated by PP-ribose-P[9]. We have also found that the electrophoretic mobilities of HGPRT from RJK44 and 47 are the same as the normal enzyme, but the enzyme from RJK3 has a reduced mobility which allows it to be easily separated from the wild-type enzyme. The demonstration of kinetic and electrophoretic alterations of HGPRT from our 8-azaguanine resistant variants provided evidence that they were, in fact, HGPRT mutants and again demonstrated the heterogeneity in our cell collection. Under normal conditions, these mutant cells are functionally HGPRT-deficient. That is, they do not incorporate hypoxanthine and are resistant to 8-azaguanine. But in the presence of aminopterin, which is known to increase the intracellular concentration of PP-ribose-P[11], they become HGPRT proficient. Thus, the basis of their complex phenotypes seems to be regulation of the mutant forms of the enzyme by changes in substrate concentrations which do not affect the normal enzyme.

We have also obtained evidence for structural gene mutations by analyzing clones with no detectable HGPRT activity. This has been done with a variety of immunological assays using sera raised against HGPRT purified from Chinese hamster brains. Beaudet et al.[12] first demonstrated the presence of material that cross-reacted with HGPRT (CRM) by showing that extracts of HGPRT-deficient mutants such as RJK39 would displace normal enzyme from immuno-precipitates while extracts from other mutants such as RJK36 would not. Both of the clones mentioned had been induced by EMS. A more informative use of the antisera was the immunoprecipitation of HGPRT from extracts of cells that had been radiolabelled with

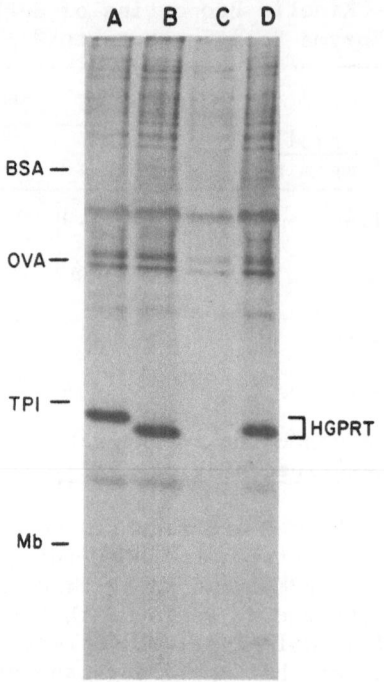

Fig. 1. Assay for the production of HGPRT antigen by mutant cell
lines. Extracts of cellular proteins labelled with 35S-methio-
nine were subjected to immunoprecipitation and the products analyzed by SDS-polyacrylamide gel electrophoresis as described[9].
Lanes A = wild-type, RJK0 cells; B = the HGPRT-positive mutant
RJK3; C = the HGPRT-negative, CRM-negative mutant RJK36; and D =
the HGPRT-negative, CRM-positive mutant RJK39.

35S-methionine and its analysis by electrophoresis in SDS-polyacrylamide gels. As shown in Figure 1, RJK39 produces normal
amounts of HGPRT protein, although the clone has no detectable
enzyme activity, but the protein cannot be found in RJK36 extracts. Furthermore, HGPRT from RJK39 migrates more rapidly during
the electrophoresis in the presence of SDS. Also shown in Figure 1
is the HGPRT immunoprecipitated from RJK3. That active but kinetically altered enzyme was also found to have an increased mobility
through SDS-gels. Whether these changes in electrophoretic mobility represent reductions in molecular weight or the presence of
amino acid substitutions that alter the amount of SDS bound to the
proteins has yet to be totally clarified but this immunological
assay has been able to identify both quantitative and qualitative
variation of HGPRT in our mutants. A complement-fixation assay
developed by C.S. Chiang[8] has provided us with similar and
confirming information. The results of our CRM investigations are

shown in Table 2. We screened 57 clones with no detectable HGPRT
activity and found only five (two EMS induced clones, RJK5 and 15,
two induced with MNNG, RJK39 and RJK45, and one induced with
ICR191E, RJK463) with detectable CRM. This is probably an under-
estimate because we have also scored four clones which will incorporate hypoxanthine as being CRM-negative. Our immunological
assays should detect clones having as little as 5% to 10% as much
enzyme as normal cells but clones synthesizing small amounts of
HGPRT or unstable forms of the enzyme can be missed. An example of
this will be described later.

During our immunological investigations we discovered that none
of our various sera would recognize the subunit protein of HGPRT if
we used agents such as urea or heat to disrupt the multimeric
structure of the active enzyme. Since such sera would not detect
synthesis of HGPRT subunits that fail to associate into multimers
or the fragments of HGPRT which would be extected to result if the
mutants carried nonsense or frameshift mutations, we raised antisera against subunit protein isolated from purified enzyme[7].
We have identified sera which will not react with the native form
of the enzyme but will immunoprecipitate HGPRT protein after enzyme
preparations or cell extracts have been treated with 8M urea.
However, reexamination of clones that were scored as CRM-negative
with previous sera has failed to identify any that contain free
subunits or fragments of HGPRT. Perhaps such molecules would be
too short-lived for detection by these assays.

We have recently used the combination of immunoprecipitation
and SDS-polyacrylamide gel electrophoresis to purify HGPRT radiolabelled with various amino acids. Tryptic digests of those preparations have been analyzed by HPLC to develop peptide maps for the
proteins from normal and mutant clones[13]. For the wild-type
enzyme we have identified eight tryptic peptides that terminate in
arginine, ten which end with lysine and one which has neither of
those amino acids but does contain a methionine residue. That last
peptide has been identified as the carboxyl terminus of the protein
by a variety of assays[13]. Analysis of the enzyme from the
mutant RJK3 described above has demonstrated that its most basic
lysine-containing peptide is altered in that its elution from the
HPLC column is delayed. Its arginine-terminated peptides and carboxyl terminus are indistinguishable from those of wild-type enzyme. Thus, the mutation that altered the amino acid content of
this single peptide has changed the kinetics of the enzyme as well
as its electrophoretic mobility under native and denaturing conditions. HGPRT from the enzyme-negative but CRM-positive line RJK39
also has only one peptide alteration. Lysine peptide number 8 is
absent from the elution profiles of its tryptic peptides. It might
either coelute with another peptide or be insoluble in our solvent
systems.

Table 4. Characterization of Mutants Producing
          Thermosensitive HGPRT[a]

Clone	Growth Temperature	APRT nmol·mg^{-1}·min^{-1}	HGPRT nmol·mg^{-1}·min^{-1}	$t_{\frac{1}{2}}$[b]
RJKO	33°	3.0	3.4	135
	39°	2.2	2.6	
RJK526	33°	4.4	0.09	15
	39°	2.5	0.001	
RJK531	33°	4.0	0.28	20
	39°	2.1	0.03	

a Enzyme assays were conducted on extracts of wild-type, RJKO, or
mutant cells grown at 33° or 39° as described by Fenwick and
Caskey[14].

b Half-life or HGPT in cell extracts incubated at 39°.

In addition to the general procedures described above, we have
also carried out selective protocols to isolate thermosensitive
mutants of HGPRT. This was done by exposing nonmutagenized cells
to 6-thioguanine at 39° and isolating colonies that continued to
grow when the temperature was reduced to 33° and the medium
changed to HAT. As illustrated in Table 4, the specific activity
of HGPRT is decreased in those cells, even when they are grown at
the permissive temperature of 33° and it is more severely decreased in extracts prepared from cells grown in non-selective
medium at 39°. Thermosensitivity of the altered forms of HGPRT
from the mutants can also be demonstrated in vitro by incubating
extracts of cells grown at the permissive temperature. The half-
lives at 39° of HGPRT from RJK526 and 531 are 7- to 9-fold
shorter than that of the normal enzyme.

One of the most impressive aspects about the studies to characterize the 8-azaguanine resistant clones in our collection has been
the heterogeneity of the observed phenotypes. There are enzyme-
positive mutants in which the HGPRT has kinetic alterations but for
those studied thoroughly each line can be distinguished by its
particular alterations. Furthermore, they can be subdivided by the
presence or absence of electrophoretic variation. Amongst the
enzyme-negative mutants are CRM-positive clones which again can be
subdivided by electrophoretic variation of the protein under denaturing conditions. This indicates that we have selected clones
carrying many different mutations. We have also noted that use of
mutagens alters the types of mutants isolated. Thus, further analysis of the clones should allow us to draw conclusions on the ac-
tions of specific mutagens in animal cells. Of particular

interest will be determination of the spontaneous events that lead to HGPRT deficiency. To this point, we have concentrated on the clones producing altered forms of HGPRT since they provide evidence of structural gene mutations. However, it is likely that some of the HGPRT-negative, CRM-negative clones might carry mutations or major DNA alterations that alter regulation or expression of the gene. Our first step to analyze those clones has been to subject them to reversion studies.

REVERSION OF HGPRT MUTANTS

Many of the HGPRT mutants that we have characterized are suitable for reversion analysis. This has been done by plating the cells at a variety of densities in non-selective medium with or without prior treatment with mutagens. After one or two days the medium is changed to HAT and colonies which form in that counter-selective medium are isolated for phenotypic characterization. Of course, most of those mutants described above which produce altered but functional HGPRT cannot be subjected to this analysis because their plating efficiencies are not reduced in HAT medium. Table 5 summarizes the results of our experience with reversion of our HGPRT-deficient mutants. Of 21 mutants tested we have successfully reverted 18 to the HGPRT-positive state. The three mutants which have failed to revert, RJK33, 36 and 43 were all induced by EMS and are CRM-negative as well as HGPRT-negative. We noted above that a number of the EMS-induced mutants have properties consistent with the presence of missense mutations. Since we have not been able to revert others of that group, however, we must entertain the possibility that EMS mutagenesis has also resulted in more severe DNA alterations. We also noted above that none of the spontaneous or

Table 5.   Reversion of HGPRT Mutants [a]

Forward	Number of Clones	
Mutagen	Tested	Reverted
None	4	4
MNNG	7	7
EMS	8	5
UV	2	2
	21	18

[a]HGPRT-negative mutants induced with the listed mutagens were subjected to reversion analysis with and without prior mutagen treatment and clones that would give rise to HGPRT-positive isolates were identified as described previously[6,15].

Table 6. Mutagen Stimulation of Reversion [a]

Clone	Forward Mutagen	Revertant Frequency			
		Spontaneous	MNNG	EMS	UV
RJK10	MNNG	9×10^{-7}	4×10^{-5}	2×10^{-5}	4×10^{-5}
RJK37	EMS	$< 2 \times 10^{-7}$			3×10^{-6}
RJK39	EMS	$< 2 \times 10^{-7}$	2×10^{-6}	8×10^{-7}	

[a]Experimental procedures and calculations of revertant frequencies have been described previously[6,15].

UV-induced mutants had detectable HGPRT activity or CRM, yet we show in Table 5 that all six of the mutants tested from those categories were able to revert. Thus the mutations responsible for their phenotypes cannot be major deletions. The same preliminary conclusion can be drawn about the seven MNNG-induced mutants tested since they all reverted.

All of the mutants we have analyzed have been quite stable. That is, they have very low spontaneous reversion frequencies. Of the three examples listed in Table 6, only RJK10 has given rise to spontaneous revertants under our experimental conditions. However, revertants have been isolated from the others after treatment with mutagens.

One of our most extensive reversion analyses has been conducted with RJK39[15]. We noted above that this HGPRT-negative but CRM-positive clone carries a mutation which slightly increases the mobility of the HGPRT protein during SDS-polyacrylamide gel electrophoresis and removes one lysine-containing tryptic peptide from the HPLC elution pattern. Of the 22 independently isolated revertants of RJK39 that we have tested, the phenotypic properties of ten as well as the electrophoretic and enzymatic properties of their HGPRTs were indistinguishable from those of wild-type cells. Thus they might be true revertants and carry actual reversals of the original mutation. Although the electrophoretic mobilities of HGPRT from 11 of the remaining 12 revertants returned to the normal pattern, hypoxanthine incorporation by cells of those clones was stimulated by aminopterin and the cells were resistant to 8-azaguanine indicating that they produced kinetically altered HGPRT. One clone regained enzyme activity but its HGPRT maintained the electrophoretic alteration noted for RJK39. Thus, 12 of the revertants must carry second-site mutations in the HGPRT locus which restore enzyme activity by intragenic suppression.

We have analyzed the tryptic peptides of HGPRT from three of the revertants of RJK39 with special attention being paid to lysine

peptide number eight, which is absent in the profile of the mutant's protein[13]. That peptide reappeared in the peptide profiles of all three revertants. Its elution position was normal for
the proteins prepared from RJK631 and 635. The HGPRT phenotype of
the latter is indistinguishable from that of wild-type cells while
the phenotype former indicates enzymatic abnormalities of its HGPRT
(i.e., it is resistant to 8-azaguanine and aminopterin stimulates
its incorporation of hypoxanthine). Lysine peptide number eight
from HGPRT of a second revertant with the altered phenotype,
RJK634, elutes earlier from the HPLC column than its normal counterpart which suggests that the peptide contains an amino acid
substitution rather than a restoration of the normal amino acid
sequence. The heterogeneity observed among the phenotypes of the
RJK39 revertants argues that several different genetic alterations
are able to restore the loss of function generated by the original
mutation.

By analyzing the revertants of RJK39 we have also been able to
gather information on the X chromosome of Chinese hamsters. One
revertant, RJK630, was found to be pseudo-tetraploid and electrophoresis of HGPRT from that clone revealed that it produced the
mutant form of the protein as well as HGPRT having the wild-type
mobility[15]. Thus, the line was heterozygous for HGPRT and
presumably carried two active chromosomes. Karyotypic studies
revealed that the entire long arm of the X chromosome in RJK39 had
been deleted or translocated to an autosome and that RJK631 had two
copies of the short arm. We isolated HGPRT-deficient segregants of
RJK630 and found that they had only one short arm of X. In addition they produced only the mutant form of HGPRT and had half as
much activity for phosphoglycerate kinase and glucose-6-phosphate
dehydrogenase (G6PD) as did RJK630. These studies confirmed the
localization of the HGPRT gene to the short arm of the Chinese
hamster X[16] and mapped the two other X-linked enzymes to that
arm.

Analysis of the UV-induced revertants of RJK37 has also revealed evidence that second-site mutation or intragenic suppression
is a common mechanism for phenotypic reversion of our mutants.
RJK37 is a HGPRT-deficient and CRM-negative mutant which was induced by EMS. It will grow in the presence of 8-azaguanine or
6-thioguanine but not HAT. Two of its HAT-resistant revertants,
RJK255 and RJK256, will not grow in medium containing 6-thioguanine
but they are not affected by 8-azaguanine. Furthermore, hypoxanthine incorporation by the revertants is low unless aminopterin is
present in which case it is stimulated at least 30-fold. As would
be expected, given those findings, the HGPRTs from RJK255 and
RJK256 were found to have altered kinetic properties. As shown in
Figure 2, those enzymes have reduced affinities for PP-ribose-P and
are activated by that substrate. Thus they are similar to the
forward mutants such as RJK3 described above. However, the km for

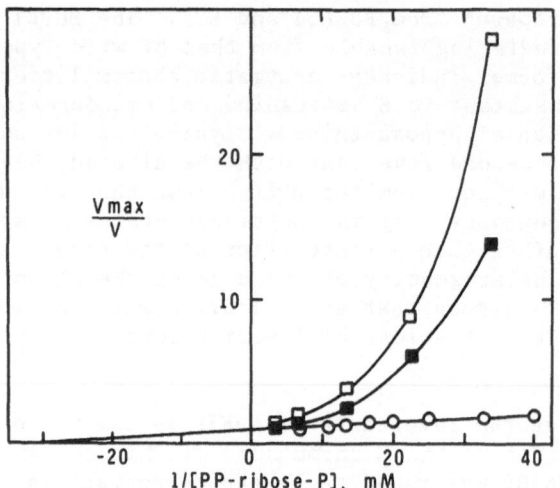

Fig. 2. Kinetic analysis of HGPRT from revertants of RJK37. Rates
of product formation were determined using extracts of wild-type
(o), RJK255 (□), and RJK256 (■) cells as described[9].

hypoxanthine is also altered for the HGPRT from RJK255 and RJK256.
As shown in Figure 3, the values derived for both the revertants
(about 33μM) are about 15-fold higher than that of the normal
enzyme (2μM). The latter property distinguishes these enzymes
from those of the forward mutants. In addition, the production of
altered enzymes by such revertants provides information about the
mutation carried by their parent strain. Since RJK37 has reverted,
we know that it cannot carry a deletion that involves structural
portions of the HGPRT. Being a CRM-negative mutant, it could be
affected in either structural or regulatory regions of the locus.
As has been concluded by other authors[17], however, synthesis of
altered enzymes in the revertants indicates that the original mutation in RJK37 must alter the coding sequence of the gene and that
the genetic alteration acquired by the revertants compensates for
rather than reverses the original.

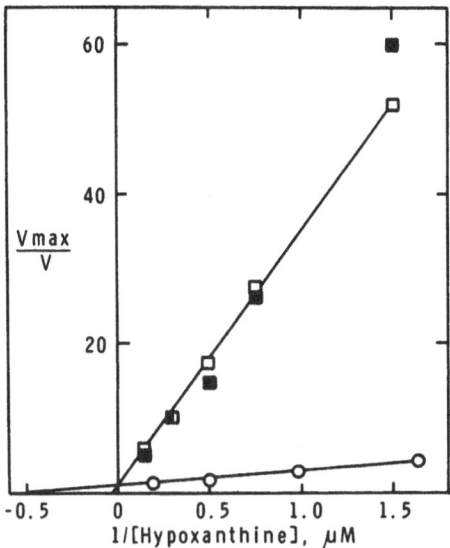

Fig. 3. Kinetic analysis of HGPRT from revertants of RJK37. Rates of product formation were determined using extracts of wild-type (○), RJK255 (□), and RJK256 (■) cells as described[9].

Analysis of HGPRT from revertants of the enzyme-deficient, CRM-negative mutant RJK10 has provided evidence of a unique type of enzyme alteration[18]. As was shown in Table 6, we have been able to isolate spontaneous revertants of this mutant but mutagenesis with a variety of agents substantially increased the frequency of HAT-resistant cells in the population surviving the treatment. HGPRT activity was detectable in extracts of all 14 independent isolates tested, but the specific activities for the various clones ranged from less than 10% to more than 60% of the wild-type value. Using a quantitative immunoprecipitation test, we found that HGPRTs from all of the revertants of RJK10 are less reactive with anti-HGPRT serum than enzyme from wild-type cells, revertants of other mutants, or enzyme positive mutants. Furthermore, enzymes from a RJK10 revertant and wild-type cells are immunoprecipitated independently from mixtures of the two cell extracts. Thus, one or more of the antigenic determinants present on Chinese hamster HGPRT (those of highest affinity) are either missing or present in altered form on HGPRT from revertants of RJK10. As discussed above, this indicates that RJK10 carries a mutation in the structural gene for HGPRT and that secondary mutations in the gene give rise to the revertants which produce the antigenically altered enzymes.

As illustrated in Table 7, various revertants of RJK10 have vastly different plating efficiencies in medium containing 6-thioguanine, the selective agent for HGPRT deficiency. We have cited several instances where the cause of such behavior by mutant

Table 7. Plating efficiencies of RJK10 revertants
in 6-thioguanine

Clone	Plating Efficiency[a]
RJK159	.043
RJK160	.00053
RJK163	.021
RJK164	.0016
RJK165	.0029
RJK167	.009
RJK169	.00056
RJK170	.11
RJK171	.003
RJK172	.02
RJK173	.00043
RJK174	.003

[a]Revertant clones maintained in HAT medium were plated in
non-selective medium at various densities and 24 hr later the
medium was changed to non-selective medium +/- 6-thioguanine.
Plating efficiency represents 6-thioguanine resistant colonies/
total colony forming units.

or revertant cell lines has been found to be production of kinetically abnormal forms of HGPRT. However, this does not seem to be
the case for the revertants of RJK10. We have done our most extensive studies with RJK159 and found its HGPRT-positive phenotype to
be very unstable. As shown in Table 7, when a population of RJK159
cells that has been maintained in HAT medium is plated at low densities in non-selective medium and changed to medium containing
6-thioguanine 24 hr later, about 5% of the colony forming units
present are resistant to 6-thioguanine. When we isolated and
tested such colonies we found that they were in fact HGPRT-
negative. We have also formed somatic cell hybrids between RJK159
which are positive for glucose-6-phosphate dehydrogenase (G6PD) and
a Chinese hamster ovary cell line that is negative for both HGPRT
and G6PD. HGPRT-negative segregants from such hybrids all remain
G6PD-positive whereas segregants from hybrids formed between our
wild-type clone, RJK0, and the CHO parent are all G6PD-negative.
Thus, the HGPRT-positive phenotype of RJK159 is unstable and it
does not co-segregate with another X-linked marker (G6PD) from
hybrid cells. One can construct a number of hypotheses for the

unstable phenotype of revertants such as RJK159. The HGPRT-positive state could be dependent on an extragenic suppressor mutation that might be subject to segregation from either diploid or hybrid cells. Alternatively, the functional HGPRT allele(s) could be extrachromosomal DNA fragments, possibly generated by some sort of gene amplification, as described by others[19], which would be subject to loss at cell division. However, we know that the HGPRT-negative derivatives of RJK159 retain at least one copy of the locus because we have been able to revert them at low frequencies to HGPRT-positive clones. A complete understanding of the phenotype of these particular revertants awaits DNA sequence analysis utilizing the probes and methods described below.

Information gained from our reversion analyses has increased our understanding of the genetic alterations carried by our collection of HGPRT-negative mutants. We have been able to revert the vast majority of the clones, even those which are CRM-negative. This would indicate a low frequency of deletion mutants. We have shown that revertants of two CRM-negative mutants produce different altered forms of HGPRT. As has been argued, this provides evidence that the structural gene is altered in those mutants, but it also shows that different forward mutations are present in the two mutants. This gives us further documentation of the genetic heterogeneity in our mutant collection. The three mutants which we have been unable to revert may carry major DNA alterations in the HGPRT locus. On the other hand, they were all induced with the point mutagen EMS which often gives rise to base substitutions in other systems. Since we have found that many reversion events in other clones apparently involve second-site mutations and intragenic suppression, it is possible that the non-reverting mutants actually carry base substitutions at sites where only a true reversion will restore enzyme activity or expression. The target for such an event might be much smaller than that for second-site reversion and thus the frequency of reversion might have been beyond the limit of detection in our experiments. Productive analysis of these clones and other mutants in the collection is dependent on techniques to study transcription and DNA sequence of the HGPRT locus. The development of the recombinant DNA probes for those studies is described in the next section.

CLONING MOUSE HGPRT cDNA

To date, our analysis of the HGPRT locus has been limited to studies involving somatic cell genetics or characterizations of the altered forms of the enzyme as described above. The next logical step is to use recombinant DNA technology to analyze the mutants and revertants. However, the isolation of cDNA clones using HGPRT mRNA from Chinese hamster tissues or somatic cells has proven to be difficult because it is only .01 to .1% of the total mRNA.

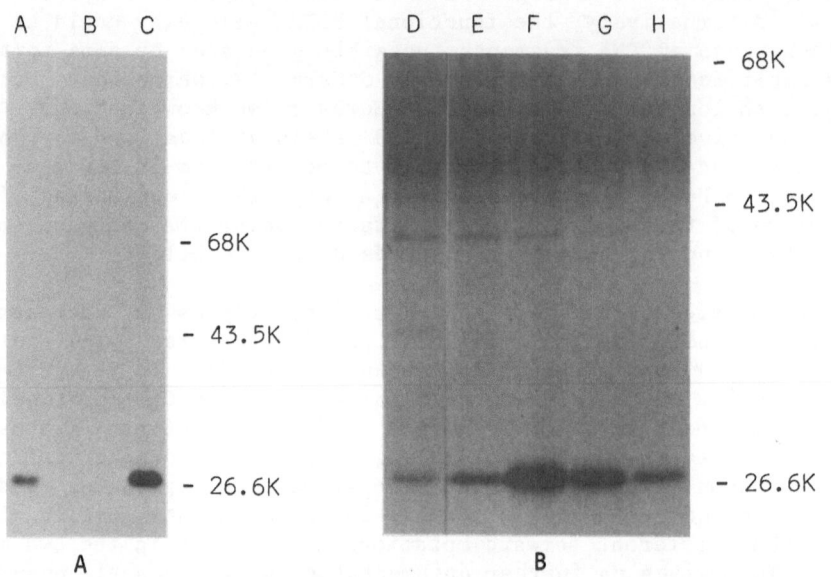

Fig. 4. Levels of HGPRT protein and mRNA in mouse neuroblastoma cells. HGPRT radiolabeled in vivo of in vitro with [^{35}S]methionine was purified by immunoadsorption with an antibody to Chinese hamster enzyme and analyzed by autoradiography after electrophoresis through SDS-polyacrylamide gels. A, HGPRT synthesized in vivo and purified from cell extracts containing equivalent amounts of acid-insoluble radioactivity (Lanes A = NB$^+$, B = NB$^-$, and C = NBR4). B, HGPRT synthesized by in vitro translation in rabbit reticulocyte lysates of oligo (dT) purified mRNA from mouse neuroblastoma cells (Lanes D = 3.7 µg NB$^-$, E = 4.6 µg NB$^+$, and F-H = .06, .03 and .015 µg NBR4). Mobilities of protein molecular weight standards are shown.

Fortunately, Melton[20] has recently isolated clones of mouse neuroblastoma cells which overproduce HGPRT and its mRNA to an extent that makes cDNA isolation possible. He isolated two classes of HAT-resistant revertants, both at frequencies of 2 x 10^{-6}, following UV mutagenesis of the NB$^-$ clone which is 6-thioguanine

resistant and has no detectable HGPRT activity. The first class
contained an HGPRT protein with the same electrophoretic mobility
and heat stability as wild-type mouse neuroblastoma (NB+) enzyme. Members of the second class all contained a variant protein
with reduced electrophoretic mobility and extreme heat lability
relative to wild-type enzyme. Although HGPRT activities in the
first class of revertants were indistinguishable from wild-type
levels, enzyme activities are much lower in all but one of the
revertants with variant protein (NBR4). In vivo determinations
indicated that the amount of immunologically detectable HGPRT protein present in NBR4 cells was 5- to 10-fold elevated over the
level in both wild-type cells (Figure 4A) and the other revertants
with variant protein. HGPRT extracted from NBR4 had twenty-fold
lower affinity for the substrate PP-ribose-P than wild-type enzyme. The low activity in other revertants with variant protein
made kinetic constant determination impractical. As described by
Melton et al.[21], using in vitro translation studies we have
found that the elevated HGPRT levels in NBR4 cells were a consequence of protein overproduction rather than a decreased rate of
HGPRT degradation. mRNA from NBR4 directed the synthesis of 25- to
50-times more HGPRT protein than equivalent amounts of wild-type
mRNA (Figure 4B). Although in vivo studies failed to detect HGPRT
protein in NB⁻ cells, its presence was detected following translation of NB⁻ mRNA. These observations are consistent with the
following model: NB⁻ cells produce normal levels of the same
variant protein found in NBR4. The cell line is 6-thioguanine
resistant because the enzyme is unstable in vivo (therefore unde-
tectable) and has poor catalytic ability. Greater stability of the
variant protein in vitro renders it detectable upon translation of
NB⁻ mRNA and explains the discrepancy between in vivo and in
vitro determinations of HGPRT overproduction in NBR4. Thus, reversion may occur by correction of the original lesion in NB⁻ producing revertants of wild-type phenotype, or by protein overproduction. NBR4 achieves near normal HGPRT enzyme activity by producing
elevated levels of the variant protein. We predict that the other
revertants with variant enzyme are also overproducers, but at a
lower level than NBR4.

As described by Brennand et al.[22], we constructed a cDNA
bank in the plasmid vector pBR322 by reverse transcription of NBR4
mRNA and G-C "tailing." The bank was first screened for clones
carrying NBR4-specific sequences by differential colony hybridization with NBR4 versus NB⁻ radioactive cDNA. From clones located
by that assay, we used a positive in vitro translation study to
identify HGPRT cDNA clones. The restriction maps of two isolates,
pHPT1 and pHPT2, are shown in Figure 5. These plasmids were used
to investigate HGPRT mRNA levels and gene copy number in NBR4
cells. RNA blot hybridization showed that an increased HGPRT mRNA
level was the basis for HGPRT protein overproduction (Figure 6A).
NBR4 cells were estimated to contain at least twenty-fold more

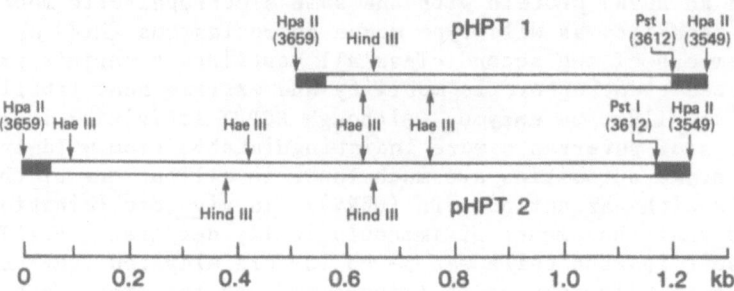

Fig. 5. Restriction endonuclease cleavage sites in recombinant plasmids pHPT1 and pHPT2. The HGPRT cDNA inserts are represented by open bars and pBR322 sequences by solid bars with numbers defining the location of the cDNA inserts in the vector molecule.

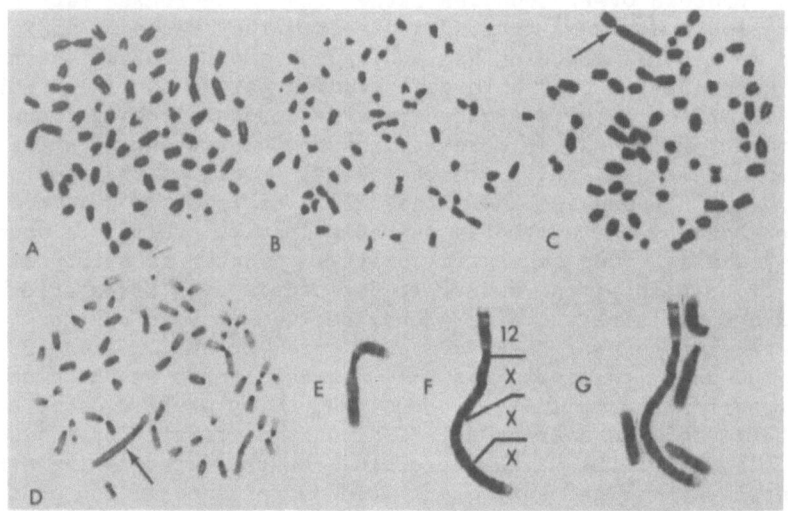

Fig. 6. Chromosome analysis of mouse neuroblastoma cells. A, Giemsa stained metaphase of NB⁺; B, Giemsa stained metaphase of NB⁻; C, Giemsa stained and D, C-banded metaphase of NBR4 with elongated marker chromosome arrowed; E, G-banded marker 3 from NB⁺, comprising a centric fusion of chromosome 12 and the X; F, G-banded elongated marker of NBR4. The p arm corresponds to chromosome 12, while the q arm consists of three copies of the X with the middle segment inverted. G, reconstruction of the NBR4 marker shown in Fig. 3F; the X; 12 translocation from NB⁺ (Fig. 3E) has been cut at the centromere and three copies of the q arm are aligned to demonstrate homology with the NBR4 marker.

HGPRT mRNA than NB⁺ or NB⁻. Southern analysis of <u>Hind</u> III
digested NB⁺ and NBR4 DNA with pHPT2, a recombinant plasmid containing 70% of the estimated full length of the HGPRT mRNA, identified four major restriction fragments in each DNA (Figure 6B).
Each sequence was amplified in NBR4 DNA; the extent of amplification was estimated to be fifty-fold.

Our cytogenetic analysis of the various neuroblastoma clones
has revealed that all revertants with variant HGPRT had a large
biarmed marker chromosome which was present in neither NB⁺, NB⁻
nor revertants with the wild-type phenotype. Only the marker chromosome from NBR4 cells has been studied in detail (Figure 7). This
chromosome, which had a single C-band located at the centromere,
was shown by G-banding to be partially homologous with the third
largest biarmed chromosome of NB⁺ and NB⁻, which is the product
of a centric fusion between chromosomes 12 and the X. The p arm of
the NBR4 marker is homologous to chromosome 12, while the q arm

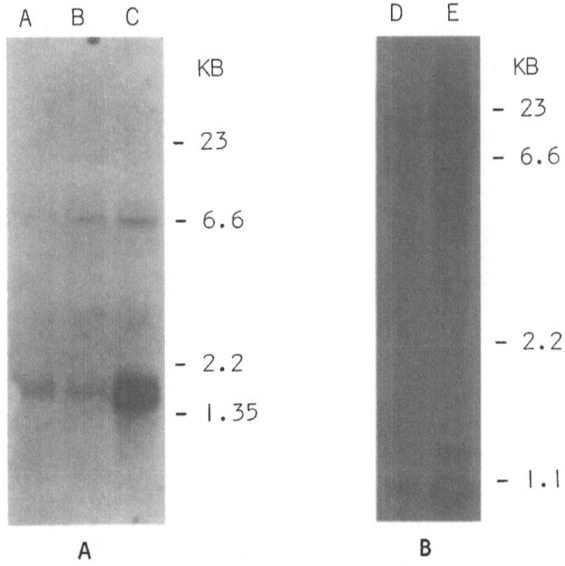

Fig. 7. Hybridization analysis of HGPRT mRNA or DNA sequences
after electrophoretic fractionation through agarose gels. After
transfer to nitrocellulose sheets the HGPRT sequences were localized by hybridization to radiolabeled cDNA recombinants and autoradiography. Positions of molecular weight standards are shown.
A, 4 μg of glyoxylated mRNA from A = NB⁻, B = NB⁺, and C =
NBR4 cells probed with pHPT1. B, DNA from D = NB⁺ (20 μg) and
E = NBR4 (10 μg) after digestion with <u>Hind</u> III probed with pHPT2.

consists of three copies of the X chromosome (less two centro-
meres), the middle segment being inverted relative to the other
two.   There was no evidence for either homogeneously staining
regions (HSRs) or double minute chromosomes in the NBR4 karyotype.

    To evaluate the potential of using the cDNA clones of mouse
HGPRT for the analysis of our collection of mutant and revertant
Chinese hamster cells or other mammalian HGPRT sequences, we have
conducted blot hybridization tests using the cDNA insert isolated
from pHPT2 after digestion with Hpa II as a probe. Figure 8A
demonstrates that this probe hybridizes to mRNAs about 1600 nucleotides long from Chinese hamster, baboon, and mouse. In addition,
it recognized a higher molecular weight molecule in total mRNA
preparations from NBR4 that was not found in the sucrose gradient
fraction of NBR4 mRNA maximal for in vitro HGPRT translation (lane
C). This larger molecule may be a precursor to the 1600 base

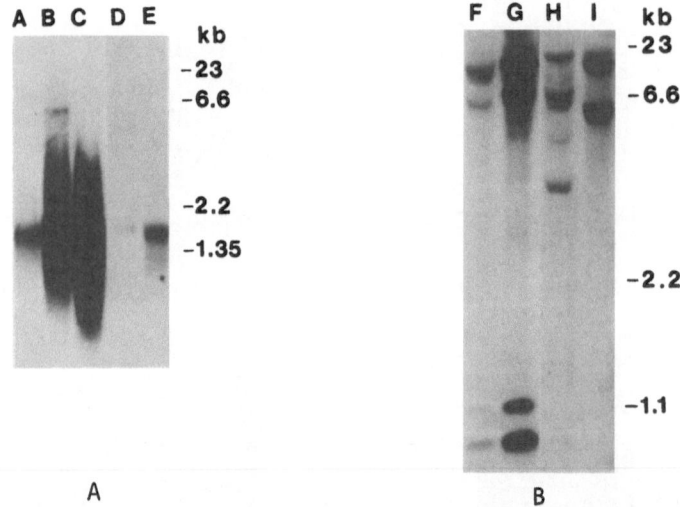

Fig. 8.   Recognition of HGPRT sequences from other mammals by the
mouse HGPRT cDNA clone.   Analysis of HGPRT sequences was conducted
as described in Fig. 7 except that the cDNA insert from pHPT2,
isolated after Hpa II digestion, served as the probe. A, characterization of mRNA purified from A = NB+ (10 µg), B = NBR4
(10 µg total mRNA),   C = NBR4 (2 µg after sucrose gradient
purification), D = baboon brain (20 µg), and E = Chinese hamster
cells (20 µg).   Autoradiography was 20 hr for lanes A to C and 70
hr from D and E.   B, results after digestion with Hind III of DNA
isolated from F = NB+ (20 µg), G = NBR4 (10 µg), H = human
lymphocytes (20 µg), and I = Chinese hamster cells (20 µg).

mRNA. The results obtained when the pHPT2 probe was hybridized to
Hind III-digested mouse, Chinese hamster, and human lymphocyte DNAs
are shown in Figure 8B. For mouse DNA, the approximate sizes of
the recognized fragments are 13 kb, 9.3 kb, 6.4 kb, 1.1 kb and 0.9
kb. The banding patterns obtained with DNA from other species are
different from those of mouse. Only 3 bands (17.5 kb, 11.3 kb and
5.5 kb) are visible in Chinese hamster (lane I) and 5 bands (17.5
kb, 6.5 kb, 6.0 kb, 4.6 kb and 3.5 kb) in human (lane H). Furthermore, the intensities of the signals are similar for NB+ (mouse),
RJKO (hamster) and human DNAs, demonstrating the ability of this
probe to detect single-copy HGPRT sequences in a wide variety of
mammalian DNAs.

FUTURE DIRECTIONS

     Our establishment of a large collection of clones carrying
mutations which affect the HGPRT gene and the subsequent isolation
of recombinant DNA molecules that can be used to identify HGPRT
sequences in nucleic acids now make possible investigations to
characterize the structure and expression of the HGPRT locus in
normal and mutant mammalian DNAs. We will be able to examine mRNA
populations from the CRM-negative mutants to determine whether any
of the mutations they carry block transcription of the gene. By
analyzing the fragmentation of the locus with restriction endonucleases, we can look for the presence of deletions, insertions, and
other alterations that might change the relationships of restriction sites. Even though our cDNA clones are from mouse, our
results indicate that they can be used at these levels to study the
gene in the mutant Chinese hamster cells or cells obtained from
families affected by the Lesch-Nyhan syndrome. Fine-structure
analysis of our mutants at the level of nucleotide sequence will of
course require isolation of recombinant DNAs from the Chinese
hamster or human sources, but the mouse probe will facilitate those
operations. To us it is very exciting to know that we will now be
able to compare and contrast spontaneous mutations and those induced with a variety of mutagens in mammalian cells, or to correlate specific alterations in the physical or enzymatic properties
of HGPRT with the location of mutations within the gene.

REFERENCES

1. C.T. Caskey and G.D. Kruh, The HPRT locus, Cell 16:1 (1979).

2. W. Szybalski, Resistance to 8-azaguanine, a selective marker
   for a human cell line, Microbiol. Genet. Bull. 16:30 (1958).

3. W. Szybalski and E.H. Szybalska, Drug sensitivity as a genetic
   marker for human cell lines, U. Mich. Med. Bull. 28:277 (1962).

4. J.E. Seegmiller, F.M. Rosenbloom and W.M. Kelley, Enzyme defect associated with a sex-linked human neurological disorder and excessive purine synthesis, Science 155:1682 (1967).

5. D.K. Ford and G. Yerganian, Observations on the chromosomes of Chinese hamster cells in tissue culture, J. Natl. Cancer Inst. 21:393 (1958).

6. F.D. Gillin, D.J. Roufa, A.L. Beaudet and C.T. Caskey, 8-azaguanine resistance in mammalian cells I. Hypoxanthine-guanine phosphoribosyltransferase, Genetics 72:239 (1972).

7. C.T. Caskey, R.G. Fenwick, G. Kruh and D. Konecki, Characterization of premature chain termination mutants of hypoxanthine-guanine phosphoribosyltransferase and their revertants in Chinese hamster cells, in Nonsense Mutations and tRNA Suppressors, J.E. Celis and J.D. Smith, Eds., Academic Press, London (1979).

8. C.S. Chiang, Characterization of HPRT mutations in cultured Chinese hamster cells, Ph.D. Thesis, Baylor College of Medicine, Houston, Texas (1977).

9. R.G. Fenwick, Jr., T.H. Sawyer, G.D. Kruh, K.H. Astrin and C.T. Caskey, Forward and reverse mutations affecting the kinetics and apparent molecular weight of mammalian HGPRT, Cell 12:383 (1977).

10. R.G. Fenwick, Jr., Virazole as an activator of mutant HGPRT and a selective agent for adenosine kinase mutants, Abstracts – International Congress of Biochemistry, Toronto, Canada (1979).

11. P.J. Benke and D. Dittmar, Phosphoribosylpyrophosphate synthesis in cultured human cells, Science 198:1171 (1977).

12. A.L. Beaudet, D.J. Roufa and C.T. Caskey, Mutations affecting the structure of hypoxanthine-guanine phosphoribosyltransferase in cultured Chinese hamster cells, Proc. Natl. Acad. Sci. USA 70:320 (1973).

13. G.D. Kruh, R.G. Fenwick, Jr. and C.T. Caskey, Structural analysis of mutant and revertant forms of Chinese hamster hypoxanthine-guanine phosphoribosyltransferase, J. Biol. Chem. 256:2878 (1981).

14. R.G. Fenwick, Jr. and C.T. Caskey, Mutant Chinese hamster cells with a thermosensitive hypoxanthine-guanine phosphoribosyltransferase, Cell 5:115 (1975).

15. R.G. Fenwick, Jr., Reversion of a mutation affecting the molecular weight of HGPRT: Intragenic suppression and localization of X-linked genes, Somatic Cell Genet. 6:477 (1980).

16. S.A. Farrell and R.G. Worton, Chromosome loss is responsible for segregation at the HPRT locus in Chinese hamster cell hybrids, Somatic Cell Genet. 3:539 (1977).

17. L.A. Chasin, A. Feldman, M. Konstam and G. Urlaub, Reversion of a Chinese hamster cell auxotrophic mutant, Proc. Natl. Acad. Sci. USA 71:718 (1974).

18. R.G. Fenwick, Jr., J.J. Wasmuth and C.T. Caskey, Mutations affecting the antigenic properties of hypoxanthine-guanine phosphoribosyltransferase in cultured Chinese hamster cells, Somatic Cell Genet. 3:207 (1977).

19. R.T. Schimke, R.J. Kaufman, F.W. Alt and R.F. Kellems, Gene amplification and drug resistance in cultured cells, Science 202:1051 (1978).

20. D.W. Melton, Cell fusion-induced mouse neuroblastoma HPRT revertants with variant enzyme and elevated HPRT protein levels, Somatic Cell Genet. 7:331 (1981).

21. D.W. Melton, D.S. Konecki, D.H. Ledbetter, J.F. Hejtmancik and C.T. Caskey, In vitro translation of HPRT mRNA: Characterization of a mouse neuroblastoma cell line with elevated HPRT protein levels, Proc. Natl. Acad. Sci. USA 78:6977 (1981).

22. J. Brennand, A.C. Chinault, D.S. Konecki, D.W. Melton and C.T. Caskey, Cloned cDNA sequences of the HPRT gene from a mouse neuroblastoma cell line found to have amplified genomic sequences, Proc. Natl. Acad. Sci. USA, in press.

TOWARDS MAPPING GENES TO CHROMOSOMES USING TWO DIMENSIONAL GEL
ELECTROPHORESIS OF PROTEINS AND SOMATIC CELL HYBRIDIZATION.
PRELIMINARY STUDIES.

Rodrigo Bravo[1], Stephen J. Fey[1], Heather
Macdonald-Bravo[1], Reinhold Shäfer[2], Klaus
Willecke[2], and Julio E. Celis[1]

[1]Division of Biostructural Chemistry, Department of
Chemistry, Aarhus University, DK-8000 Aarhus C, Den-
mark, and [2]Institut für Zellbiologie, Universität
Essen, D-4300 Essen 1, FRG

1. INTRODUCTION

Fusion of somatic cells followed by loss of chromosomes from
one of the parents represents a unique tool to assign phenotypic
traits to particular chromosomes or to regions of chromosomes[1-7]. This approach, however, is limited by the small
number of genetic markers available for the selection of hybrids[8] and more seriously by the lack of sensitive assays to
determine how much information is coexpressed in the hybrids at
the molecular level[6, 7, 9-13].

Recently, thanks to the development of two dimensional gel
electrophoresis of proteins[14, 15] and to improved [^{35}S]-
methionine labelling techniques[16-19], it has been possible to
separate and catalogue nearly 1200 acidic and basic cellular polypeptides under conditions in which their coordinates can be repro-
ducibly determined[20, 21]. These technical developments have
opened the possibility for mapping genes without previous knowledge of the function of the polypeptides for which they code.

The purpose of this article is to describe experiments which
suggest that in fact this approach may be feasible and to discuss
some problems which are now becoming evident. To keep a coherent
presentation, we will be concerned only with the possibility of
mapping mouse polypeptides, although this approach could be used
to map genes from other species. First, we shall present our
current catalogue of mouse polypeptides based on separation by two

dimensional gel electrophoresis[21]. Secondly, we will examine
polypeptide synthesis in a hamster x mouse cell hybrid that retained all the mouse chromosomes (Schäfer et al., in preparation).

2. SEPARATION AND CATALOGUING OF MOUSE FIBROBLAST POLYPEPTIDES

A total of 1147 [35S]-methionine labelled polypeptides (826
acidic [IEF] and 321 basic [NEPGHE]) from asynchronous secondary
mouse kidney fibroblasts have been separated using high resolution
two dimensional gel electrophoresis[21] (Figure 1). The polypeptides have been numbered from higher to lower molecular weight
and from the origin of the first dimension[21]. Like in HeLa
cells[20] we have selected 101 original numbers (67 in the
acidic and 34 in the basic side). Additional spots between two
original numbers (big numbers) have been indicated with letters.
In cases where there were more than 26 numbers between two original numbers we have indicated them with the letter z followed by a
small number (see also Tables 1 and 2). The coordinates (M.W. and
relative mobility respect to β-actin [IEF] or polypeptide 9
[NEPHGE]) and in a few cases the identity of some polypeptides are
given in Tables 1 and 2. We felt justified in standardizing a
number for the polypeptides in these fibroblasts as we have observed that fibroblasts obtained from several tissues present a
remarkable similar polypeptide pattern to that shown in Figure 1
(Bravo et al., in preparation).

An important proportion of the major cellular polypeptides
labelled with [35S]-methionine are clustered in a region of the
gels between MWs 35K and 70K[20, 21]. Of these, the major polypeptide is actin (13% of total protein label), which is composed
mainly of the β (IEF 47) and γ (IEF 47g) variants. Other
major polypeptides correspond to α (IEF 19 and 19c) and β-
tubulin (IEF 21); vimentin (IEF 22); three polypeptides present in
intermediate filament enriched cytoskeletons (IEF 12, 24 and
31)[22] and a tropomyosin variant (IEF 54a). IEF 51f and 52
correspond to polypeptides whose relative proportions are sensitive to changes in growth rate ([21]; Bravo et al., in preparation).

Figure 1. Two dimensional gel electrophoresis of [35S]-methio-
nine labelled polypeptides from asynchronous secondary (p4) mouse
kidney fibroblasts labelled for 20 hr. Both fluorograms (IEF and
NEPHGE) have been overexposed in order to reveal minor polypeptides. In IEF the pH ranges from 7.5 (left) to 4.5 (right). In
NEPHGE the pH varies from 7.5 (right) to 10.5 (left). Taken from
reference 27.

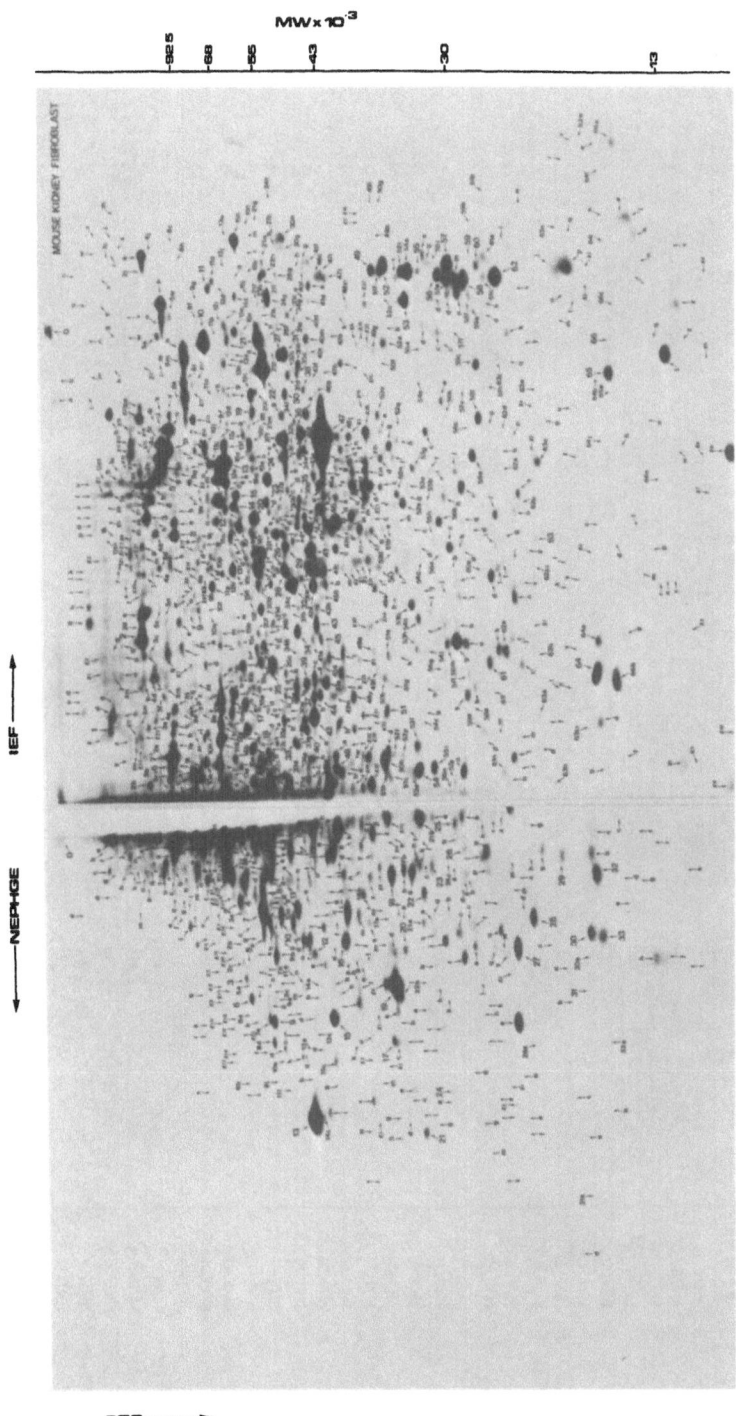

TABLE 1. Coordinates of Some Major Acidic Polypeptides

Polypeptide	Coordinates Molecular weight/ mobility relative to β-actin [a]	Other polypeptides with the same original number	Polypeptide	Coordinates Molecular weight/ mobility relative to β-actin	Other polypeptides with the same original number	Polypeptide	Coordinates Molecular weight/ mobility relative to β-actin	Other polypeptides with the same original number
0	nd/1.29	a-z8	22[b] Vimentin	54/1.18	a-h	45	43.5/0.44	a-b
1	132/0.20	a-z28	22g	54/1.22		46[c]	43.5/0.80	a-c
2	110/0.44	a-m	23	54/0.82	a-b	46[c]	43/1.44	
3	105/0.78	a-c	24[b]	53/0.88	a-h	47[d] β-actin	43/1.00	a-o
4	100/0.83	a-j	25	53/0.37	a-f	48	42/0.09	a-z35
5	96/0.89	a-x	25f[c]	52.5/1.24		48z27	38/1.36	
5p	95/1.37		26	52/0.73	a-g	49[e]	36/1.45	a-j
5t	94/0.97		27	49.5/1.22	a-g	50	35.5/0.64	a-h
5u	94/1.02		27f Skeletin	49/0.99		51[f]	35/0.24	a-g
6	95/0.10	-	28	49.5/0.63	a-c	51[f]	35/1.43	
7	95/0.12	a-d	29	49/0.70	a-c	52[f]	35/1.45	
7a	95/0.16		30	49/1.12	a-h	52t	33/1.44	a-t
8	94/0.73	a-z17	31[b]	49/1.22	a-e	53	32.5/1.36	a-h
8m	76/1.21		32	49/0.59	a-c	54	32/0.08	a-i
8n	76/1.24		33	49/0.16	a-e	55	31.5/1.44	a-j
9	74/0.52	a-b	34	49/0.46	a-k	56	31/1.43	a-n
10	72/1.26	a-q	34g[c]	47/1.32		57	30/1.46	a-v
11	69/1.41	a-p	35	45.5/0.83	a-c	58	29/1.41	a-f
12[b]	68/0.82	a-i	36	45.5/1.01	a-b	59	28/1.47	a-f
13	67/0.93	a-y	36b	46/1.20		59c	27/1.18	
14	66/0.81	a-d	37	46/0.22		60	26/1.45	a
15	65/0.78	a-h	38	45.5/1.13	a	61	25/0.37	a-n
15h	65/1.54		39	45.5/0.32	a-b	61k	24/0.41	
16	65/0.76	a-d	40	45.5/0.19	-	62	23.5/1.43	a-24
17	64/0.26	a	41	45/0.69	a-m	63	21/0.70	a-24
18	63.5/0.20	a-r	41d[c]	44/1.37	a-f	64	18/0.34	a-j
19 (α-tubulin)	58/1.07	a-c	41e	44/1.41		65	16/1.16	a
19c(α-tubulin)	58/1.12	.	42	45/0.23	a-d	66	16/1.24	a-z10
20	59/0.78	a-i	43	44/0.47	a-f			
21 (β-tubulin)	56/1.27	a-o	44	43.5/0.29	a			

[a] From Fey et al. (21) [b] Polypeptides present in cytoplast L-H-L cytoskeletons enriched in intermediate filaments [c] Vimentin degradation products

[d] γ-actin correspond to spot 47g [e] The one dimensional peptide map of this polypeptide is similar to that of IEF 52 [f] Polypeptides sensitive to changes in growth rate.

TABLE 2. Coordinates of Some Major Basic Polypeptides

Polypeptide	Coordinates Molecular weight/ mobility relative to NEPHGE 9 [a]	Other polypeptides with the same original number	Polypeptide	Coordinates Molecular weight/ mobility relative to NEPHGE 9 [a]	Other polypeptides with the same original number
0	nd/0.01	a-v	15e	46.5/0.02	
1	95/0.16	a-z5	15l	45/0.30	
1z22	68/0.24		16	45/0.66	a-z11
2	67/0.35	a-c	16v	43/1.45	
3	66/0.77	a-b	17	41/1.16	a-b
4	65/0.28	-	18	40/0.41	-
5	64/0.74	a	19	39.5/0.88	a-j
6	62/0.91	a-1	19a	39.5/1.01	
7	61/0.53	a-j	19c	39.5/1.18	
7h	59/0.10		20	37/0.69	a-k
8	56/0.21	a-e	21	35/1.43	a-b
8d	55/1.09		22	34.5/0.49	a-c
9	55/1.00	a-j	23	33.5/0.33	a-k
10	54/0.53	a-h	24	33/1.43	a-c
10d	54/1.13		25	32/0.08	a-f
11	53/0.29	a-u	25e	31/0.04	
11b	52.5/0.15		26	31/0.20	a-z5
11c	52.5/0.50		26j	29.5/0.25	
11d	52.5/0.50		27	27/0.75	a-h
11g	52/0.44		28	25/0.62	a-c
11h	52/0.33		29	23/0.37	a-e
11o	51.5/0.60		30	20.5/0.64	a
12	50/0.69	-	31	20.5/0.93	a-c
13	50/1.51	a-d	32	18.5/0.44	-
14	50/0.15	a-1	33	18/0.71	a-1
15	47/1.01	a-1	33h	16/0.76	a-1

[a] Average from several gels.

3. POLYPEPTIDE SYNTHESIS IN A MOUSE X HAMSTER HYBRID THAT
    RETAINED ALL MOUSE CHROMOSOMES

Even though a great deal of effort has been devoted to the
study of cell hybrids[23], at present we know very little about
how much information is coexpressed in the hybrids at the level of
polypeptide synthesis. Thus, prior to any mapping attempt it is
necessary to determine first if all the recognizable polypeptides
from one of the parents, in this case the mouse, are expressed in
cell hybrids that have retained copies of all mouse chromosomes.
With this in mind we have carried out a detailed analysis of the
polypeptides synthesized by a mouse x hamster cell hybrid (hybrid
20 BW-4; Schäfer et al., in preparation) produced by fusing tumorigenic hamster WCI cells with a 20 fold excess of mouse BALB/c
fibroblasts. This hybrid retained copies of all mouse chromosomes
(Table 3, Schäfer et al., in preparation) and seemed suitable for
these studies.

Figures 2 and 3 show the two dimensional maps (IEF and NEPHGE)
of polypeptides from the hamster parent (Figure 2) and of hybrid
20 BW-4 (Figure 3) labelled for 20 hr with [^{35}S]-methionine.
The spots indicated with arrows in Figure 2 correspond to some
selected hamster polypeptides that comigrate with mouse fibroblasts polypeptides (Figure 1), and in a few cases these have been
numbered using the mouse numbering system[21]. Polypeptides
indicated with arrowheads and letters correspond to a few selected
hamster specific polypeptides that do not comigrate with mouse
polypeptides. In Figure 3 mouse specific polypeptides are indicated with arrows and in many cases these have been numbered using
the mouse numbering system. Clearly, the overall polypeptide
pattern of hybrid 20 BW-4 corresponds to that of the hamster
parent, although a significant number of mouse polypeptides are
also present (arrows in Figure 3). There are, however, an important number of basic (NEPHGE 2,4,6d,15e,15j,15k,16i,16o,16p,16y,
16z4,17,19h,21b,23,26h,27a,30a,31b) and acidic (IEF 5i,27f,49,
56e,57g,57s,59c,59f,61,61c,62j,62k,62s,63h,64a,66b,66o,66q,66z9)
mouse polypeptides that are not expressed, at least at a detectable level. Since some of these polypeptides correspond to major
proteins (see for example NEPHGE 27a in Figure 1), it is unlikely
that our failure to detect them is due to the experimental
approach used. Thus, given the chromosomal constitution of this
hybrid (Table 1), it seems likely that not all the mouse genome is
expressed, at least at the translational level. Whether whole
chromosomes or only regions of them are expressed is as yet
unknown. At present it is difficult to determine the percentage
of each mouse polypeptide expressed in the hybrid relative to the
parent but preliminary data indicate for example, that about 50%
of NEPHGE 19 and 20% of IEF 24 are observed in this hybrid.

TABLE 3. Chromosomal Constitution of Hybrid 20 BW-4

		No. of chromosomes				% of metaphases containing mouse chromosomes																		
	Total no. of meta-	Hamster		Mouse																					
Hybrid	phases	Mean	Range	Mean	Range	1	2	3	4	5	6	7	8	9	10	11	12	13	14	15	16	17	18	19	x
20 BW-4	20	22	9-33*	30	14-42	65	100	70	80	70	80	50	50	75	65	30	70	80	60	50	90	80	60	100	50

*One exceptional hybrid metaphase contained 9 hamster chromosomes and 17 mouse chromosomes.

Finally, similar analysis of hybrids which have segregated mouse chromosomes[24], showed that these expressed only some of the mouse polypeptides that were detected in hybrid 20 BW-4 (unpublished observations), indicating that in principle it should be possible to map a given polypeptide to a particular chromosome using this approach. Similar studies of hybrids carrying only a single mouse chromosome are now in progress.

4. CONCLUSIONS

Even though it may be feasible in the future to map genes to chromosomes using the proposed approach there are some important principles that are just emerging, and should be considered before engaging in a lengthy project. Firstly, as exemplified by the hamster x mouse hybrid studied here, it is not a good choice to use hamster cells as a parent to map mouse polypeptides as too many hamster and mouse polypeptides are homologous as determined by one dimensional peptide mapping (unpublished observations), and comigrate in two dimensional gels. Secondly, it is clear that not all of the mouse genetic information is expressed at the translational level, at least in the hybrids we have studied. It should therefore be emphasized that a negative result has no meaning on its own, i.e. the lack of expression of a particular polypeptide in a hybrid containing only one chromosome does not necessarily prove that the gene is not on that chromosome.

Figure 2. Two dimensional gel electrophoresis (IEF, NEPHGE) of polypeptides from the hamster parent labelled for 20 hr with [^{35}S]-methionine. Polypeptides indicated with an arrow comigrate with mouse polypeptides and in a few cases they are indicated with the corresponding mouse number (see also Fig. 1). A few hamster specific polypeptides are indicated with arrowheads and letters.

Figure 3. Two dimensional gel electrophoresis (IEF, NEPHGE) of polypeptides from hybrid 20 BW-4 cells labelled for 20 hr with [^{35}S]-methionine. Arrows indicate mouse polypeptides that are expressed. When possible polypeptides have been numbered according to the mouse numbering system ([21]; see also Fig. 1).

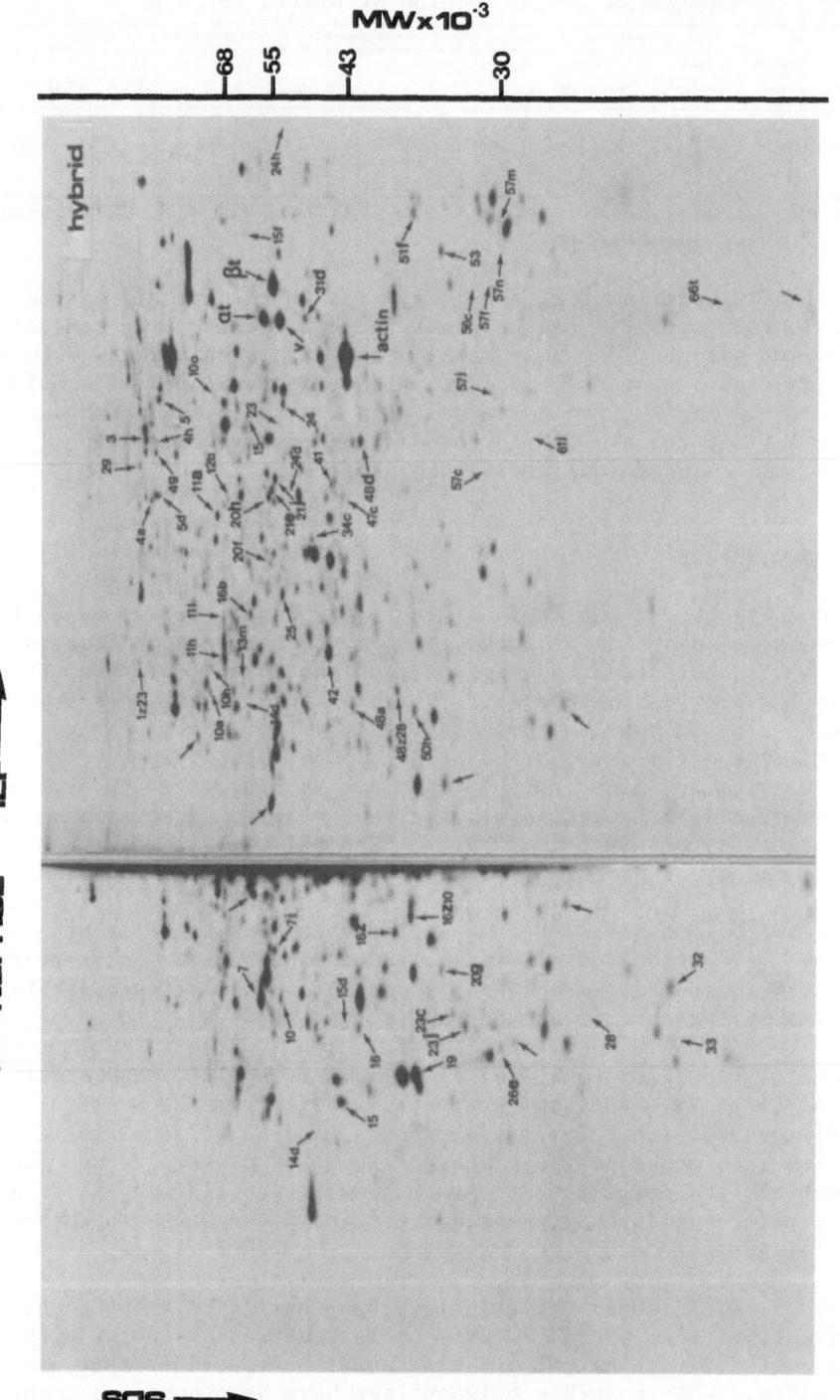

Figure 2

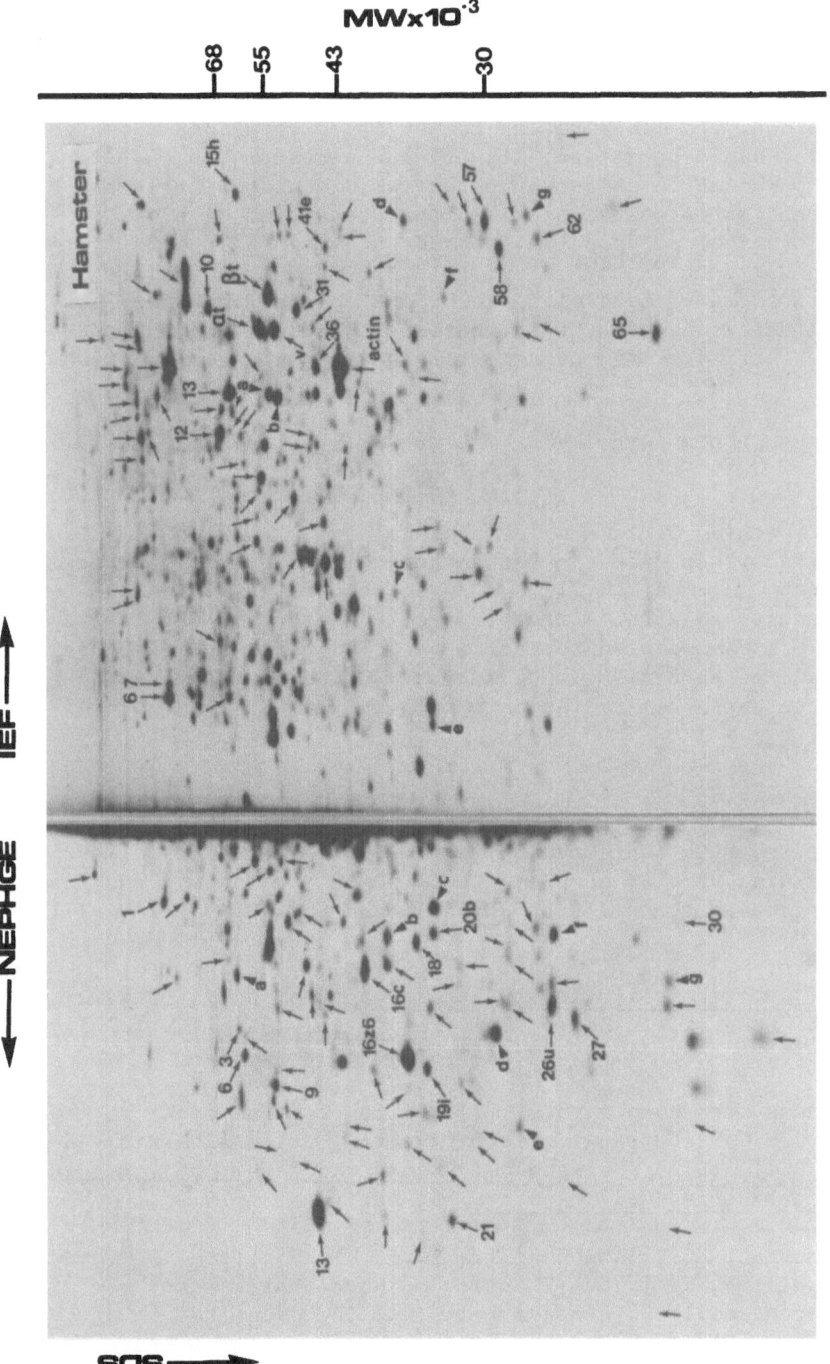

Figure 3

These drawbacks may in some cases be avoided by analysing
different hybrid combinations between unrelated species, or between diverse cell types (normal or transformed). Ideally, for
gene mapping purpose, one should study hybrids that have retained
single chromosomes or recognizable fragments of chromosomes from
one parent. To maximise the number of polypeptides whose gene
location could be mapped, it would be necessary to choose parents
whose polypeptide patterns in the two dimensional gels are as
different as is compatible with the production of hybrids. From
our library of polypeptide patterns of many cells of different
origin and species, it is evident that many proteins are conserved
both in terms of molecular weight and charge, and therefore additional assays such as two dimensional peptide analysis and monoclonal antibodies may be necessary to evaluate coexpression.

5. ACKNOWLEDGEMENTS

We would like to thank A. Celis and B. Thomsen for skilful
assistance and O. Jensen for photography. R. Bravo is a recipient
of a fellowship from the Danish Medical and Natural Science
Research Councils. S.J.F. is a recipient of an EMBO long-term
fellowship. H.M-B. is recipient of a short-term FEBS fellowship.
This work has been supported in part by grants from the Danish
Medical and Natural Science Research Councils (to J.E.C.) and the
Deutsche Forshungsgemeinschaft (SFB 102, to K.W.).

6. REFERENCES

1. Weiss, M.C. and Green, H. 1967. Human-mouse cell lines containing partial complements of human chromosomes and function-
   ing human genes. Proc. Natl. Acad. Sci. USA 58:1104-1111.

2. Goss, S.J. 1978. Gene mapping by cell fusion. In: Aspects
   of Genetic Action and Evolution (Bourne, G.H., Danielli, J.F.
   and Leon, K.W., Eds.), Int. Rev. Cytol. Suppl. 8:127-169. New
   York, San Francisco, London: Academic Press.

3. Human Gene Mapping 5 (1979): Fifth International Workshop on
   Human Gene Mapping. Birth Defects: Original Article Series
   XV, 11, 1979, The National Foundation, New York; also in Cyto-
   genetics and Cell Genetics 25, Nos. 1-4, 1979.

4. Tourian, A., Johnson, R.T., Burg, K., Nicolson, S.W. and
   Sperling, K. 1978. Transfer of human chromosomes via miniseg-
   regant cells into mouse cells and the quantitation of the
   expression of hypoxanthine phosphoribosyltransferase in the
   hybrids. J. Cell Sci. 30:193-209.

5. Ruddle, F. 1980. Gene transfer into eukaryotes. In <u>Transfer of Cell Constituents into Eukaryotic Cells</u> (Celis, J.E., Graessmann, A. & Loyter, A., Eds.) New York: Plenum Press, pp 295-310.

6. Bravo, R., Small, J.V., Celis, A., Kaltoft, K. and Celis, J.E. 1980. Tumorigenicity, actin cables and gene expression in mouse CLID x CHO cell hybrids. In <u>Transfer of Cell Constituents into Eukaryotic Cells</u> (Celis, J.E., Graessmann, A. & Loyter, A., Eds.) New York: Plenum Press, pp 275-294.

7. Bravo, R., Fey, S.J. and Celis, J.E. 1981. Gene expression in murine hybrids exhibiting different morphologies and tumorigenic properties. <u>Carcinogenesis</u>, <u>in press</u>.

8. Siminovitch, L. 1976. On the nature of hereditable variation in cultured somatic cells. <u>Cell</u> <u>7</u>:1-11.

9. Kuter, D.J. and Rodgers, A. 1975. The synthesis of ribosomal proteins and ribosomal RNA in a rat-mouse hybrid cell line. <u>Exp. Cell Res.</u> <u>91</u>:317-325.

10. Deisseroth, A., Burk, R., Picciano, D., Minna, J., Anderson, W.F. and Nienhaus, A. 1975. Hemoglobin synthesis in somatic cell hybrids: globin gene expression in hybrids between mouse erythroleukemia and human marrow cells or fibroblasts. <u>Proc. Natl. Acad. Sci. USA</u> <u>72</u>:1102-1106.

11. Rosenberg, R.N., Vance, C.K., Morrison, M., Prashad, N., Meyne, J. and Baskin, F. 1978. Differentiation of neuroblastoma, glioma and hybrid cells in culture as measured by the synthesis of specific protein species: evidence of neuroblastglioblast reciprocal genetic regulation. <u>J. Neurochem.</u> <u>30</u>:1343-1355.

12. Leinwand, L., Straie, R. and Ruddle, F.H. 1978. Phenotypic and molecular expression of albumin in rat hepatoma x L cell hybrids. <u>Expl. Cell Res.</u> <u>115</u>:261-268.

13. Allan, M. and Harrison, P. 1980. Coexpression of differentiation markers in hybrids between Friend cells and lymphoid cells and the influence of the cell shape. <u>Cell</u> <u>19</u>:437-447.

14. O'Farrell, P.H. 1975. High resolution two-dimensional electrophoresis of proteins. <u>J. Biol. Chem.</u> <u>250</u>:4007-4021

15. O'Farrell, P.Z., Goodman, H. and O'Farrell, P.H. 1977. High resolution two-dimensional electrophoresis of basic as well as acidic proteins. <u>Cell</u> <u>12</u>:1133-1142.

16. Celis, J.E., Kaltoft, K. and Bravo, R. 1980. Microinjection into somatic cells: Direct microinjection with micropipettes and PEG erythcocyte ghost mediated microinjection. In Transfer of Cell Constituents into Eukaryotic Cells (Celis, J.E., Graessmann, A. and Loyter, A., Eds.) New York: Plenum Press, pp 1-28.

17. Bravo, R. and Celis, J.E. 1980a. A search for differential polypeptide synthesis throughout the cell cycle of HeLa cells. J. Cell Biol. 84:795-802.

18. Bravo, R. and Celis, J.E. 1980b. Gene expression in normal and virally transformed mouse 3T3B and hamster BHK21 cells. Expl. Cell Res. 127:249-260.

19. Celis, J.E. and Bravo, R. 1981. Towards cataloguing human and mouse proteins. Trends in Biomedical Sciences, in press.

20. Bravo, R., Bellatin, J. and Celis, J.E. 1981. [^{35}S]-methionine labelled polypeptides from HeLa cells. Cell Biol. Int. Rep. 5:93-96.

21. Fey, S.J., Bravo, R., Mose Larsen, P., Bellatin, J. and Celis, J.E. 1981. [^{35}S]-methionine labelled polypeptides from secondary mouse kidney fibroblasts: coordinates and one dimensional peptide maps of some major polypeptides. Cell Biol. Int. Rep. 5:491-500.

22. Bravo, R., Fey, S.J., Small, J.V., Mose Larsen, P. and Celis, J.E. 1981. Coexistence of three major isoactins in a single sarcoma 180 cell. Cell 25:195-202.

23. Ringertz, R.N. and Savage, R.E. 1976. In Cell Hybrids. New York: Academic Press, pp 245-270.

24. Schäfer, R., Doehmer, J., Druge, P.M., Rademacher, I. and Willecke, K. 1981. Genetic analysis of transformed and malignant phenotypes in somatic cell hybrids between tumorigenic Chinese hamster cells and diploid mouse fibroblasts. Cancer Research 41:1214-1221.

THE USE OF MOUSE-HUMAN AND HUMAN-HUMAN HYBRIDOMAS IN HUMAN GENETICS AND IMMUNOLOGY

Carlo M. Croce, Alban Linnenbach, Thomas W. Dolby and Hilary Koprowski

The Wistar Institute of Anatomy and Biology
Philadelphia, Pennsylvania USA 19104

INTRODUCTION

In order to investigate the molecular basis of human immuno-deficiencies and to understand the immunological basis of human autoimmune diseases, it is important to be able to clone the genes for human immunoglobulin chains and the autoantibodies responsible for the pathogenesis of the autoimmune diseases respectively.

We have recently found that mouse x human hybridomas produce more human immunoglobulin chains than their human cell parents[1]. Since this is the result of an increase in production of human immunoglobulin specific mRNAs in the hybridomas[1], we have predicted that it should be possible to clone human immunoglobulin specific cDNAs that have been transcribed from partially purified immunoglobulin mRNAs derived from the mouse x human hybridomas. Therefore we have purified human immunoglobulin specific mRNAs from hybrid cells and have characterized their specific cloned cDNAs by the hybrid selection and positive translation method[2] and by DNA sequencing[3]. In addition we have used a human B cell line derived from a patient with multiple myeloma (GM1500) to obtain mutants deficient in hypoxanthine phosphoribosyltransferase (HPRT). Such mutants, named GM1500-6TG-Al 1 and -Al 2 were found to be deficient in HPRT and to die in HAT selective medium. The HPRT deficient GM1500 cells were found to secrete human IgG ($\lambda 2$, κ) as the parental GM1500 cells[1, 4]. We have then attempted to hybridize the human GM1500 HPRT mutant cells with human lymphocytes secreting specific antibodies to determine the feasibility to produce human x human hybridomas secreting human monoclonal antibodies. We have chosen to hybridize the GM1500 cells with peripheral lymphocytes derived from a patient with very high titers of antibodies against measles virus. This patient was a 19 years old female suffering from subacute sclerosing panencephalitis (SSPE), who had first developed SSPE at age 9. The disease lasted for 2 years, leading ultimately to recovery with the persistence of only some residual symptoms. At the age of 18 years, she married and she became pregnant, then in the fourth month of pregnancy SSPE recurred with great severity. Measles virus was isolated from a brain fragment obtained from the patient. Serum from her, diluted 1:10[6], bound measles infected cells.

METHODS.
    Cell lines and hybrid production. Human lymphoblastoid (GM607, GM1056, GM923) or myeloma (GM1500) cells were obtained from the Human Genetic Mutant Cell Repository (Camden, NJ). These cells were maintained in RPMI-1640 supplemented with 10% fetal calf serum under standard conditions. Normal human peripheral blood lymphocytes were obtained from healthy donors. All somatic cell hybrids were produced with mouse BALB/c P3x63Ag8 cells deficient in hypoxanthine phosphoribosyltransferase derived from the MOPC 21 plasmacytoma that secretes IgG1 κ or the nonsecreting subline[5]. Hybrid cells were selected, maintained, and characterized as to their human isotype secretion (H and L chains) as described[4] as indicated in the legend of Table 1.

    Purification of H and L mRNAs. Minimally degraded H and L chain mRNAs were prepared from all B cell lines and hybrids by the following method. Pelleted cells were lysed for 10 min in 5 volumes of cold 50 mM Tris-HCl, pH 7.4/25 mM NaCl/5 mM magnesium acetate/10 mM 2-mercaptoethanol/30% (wt/vol) sucrose containing polyvinyl sulfate at 20 μg/ml and 0.8% Nonidet P-40, followed by centrifugation at 15,000 x g for 15 min at 1°C to remove nuclei and mitochondria. The supernatant was adjusted to 1.5% NaDodSO$_4$ and immediately an equal volume of redistilled phenol/chloroform/isoamyl alcohol, 1:1:0.01 (vol/vol), saturated with 10 mM Tris-HCl, pH 7.4/1 mM EDTA/0.1 M NaCl/1.5% NaDodSO$_4$ was added. The solution was mixed for 10 min and centrifuged at 10,000 x g for 15 min at 20°C. The aqueous phase was precipitated with 3 vol of ethanol at -20°C after three extractions. The total cytoplasmic RNA was subjected to two rounds of oligo (dT)-cellulose chromatography with heating at 70°C for 5 min prior to the second round[6]. The polyadenylated RNA was then fractionated by neutral 5-25% sucrose gradient centrifugation[7]. Gradient fractions enriched in H and L chain mRNAs based on in vitro translation criteria[8] and immunoprecipitation[4] were centrifuged repeatedly until in vitro translation indicated substantial purification.

    Synthesis of double-stranded (ds) cDNA. Two to 3 μg of human μ poly (A)[+] mRNA from GM607 or a somatic cell hybrid αD5-DH11-BC11 was primed with oligo (dT) (P.L. Biochemicals) at 50 μg/ml and subjected to reverse transcription essentially as described[9], with 900-1500 units of avian myeloblastosis virus

TABLE 1

Properties of Human Immunoglobulin mRNAs and Their _in vivo_ and _in vitro_ Synthesized Polypeptides

Cell	Source *	Ig Secreted †	Quantity(ng/10⁶cells hr)	M_r X 10^{-3} H Chain In vivo	H Chain In vitro	L Chain In vivo	L Chain In vitro	mRNA $S_{20,R}$,†s H chain	L chain
M607	Lymphoblastoid	Ig Mκ	10	77	69	24	26.5	20	13
ED	C.L. Leukemia	Ig Mκ + IgDκ	14	77	69	24	26.5	20	13
M1500	Myeloma	IgG2κ	27	50	49	23	25	17	12
M1056	Lymphoblastoid	IgA2λ	6	60	58	24	26.5	18	13
M923	Lymphoblastoid	IgA1λ	2	60	N.D.	24	N.D.	N.D.	N.D.
D5 DH11 BC11	HPL x P3	μ(IgC1κ)	58	77	69	-	-	20S	-
3 D24.3	HPL x P3	μ(IgC1κ)	46	77	N.D.	-	-	N.D.	-
7-77-F7	HPL x P3	μ(IgC1κ)	32	77	69	-	-	20S	-
SK-10-2B1-C14	GM607 x P3	μ(IgC1κ)	21	77	N.D.	-	-	N.D.	-
SK-NS6-201-C7	GM607 x NP3	μ	30	77	N.D.	-	-	N.D.	-
06.2-R4-3C6	GM1500 x P3	γ2:(IgC1κ)	35	50	N.D.	-	-	N.D.	-
SK-4-2B2-C1	GM1500 x NP3	γ2:κ	40	50	N.D.	24	N.D.	N.D.	N.D.
SK-13-2A5-C8	GM1056 x P3	α2:λ(IgC1κ)	171	60	58	24	26.5	18	13
SK-12-1D2-C3	GM923 x P3	α1:λ(IgC1κ)	63	60	N.D.	N.D.	N.D.	N.D.	N.D.
3x63Ag8	Plasmacytoma	(IgC1κ)	0	57	54;50	23	25;26	17	13
P3	Nonsecreting P3	-	0	-	-	-	-	-	-

Human Ig secreted and class identification were determined by NaDodSO₄/polyacrylamide gel analysis of cell medium class-specific immunoprecipitates labeled with [³⁵S]methionine, quantitative immunofluorescence, Ouchterlony precipitin rings with class-specific antisera, radioimmunoassays with class-specific reagents, and subclass Marchalonis assays, ND, not determined.

* HPL, human peripheral lymphocytes.

† Human Ig secreted is shown in parentheses.

† Mouse Ig secreted is shown in parentheses.

† S₂₀,R was determined in neutral 5-25% sucrose gradient of twice-purified oligo(dT)-cellulose polyadenylated RNA relative to agarose gel-purified 4S and 18S rRNA.

reverse transcriptase (obtained from J. Beard, St. Petersburg, Fl)
per ml. Second-strand synthesis was carried out as described[9]
except that Escherichia coli DNA polymerase I (Boehringer Mannheim,
grade I) at 150 units/ml, additional dNTPs (to 1 mM), and 10 mM
MgCl$_2$ were found necessary to promote about 60-70% synthesis of
the second strand. The ds cDNA was extracted with chloroform and
chromatographed on Sephadex G-100 in 20 mM NaCl/2.5 mM EDTA; void
fractions were pooled and precipitated with 3 vol of ethanol.

The ds cDNA was trimmed to blunt ends with S1 nuclease (Miles)
at 100 units/ml as described[10], extracted, rechromatographed on
Sephadex G-100, and precipitated. Homopolymeric tracts of dC were
added to the 3' ends of the ds cDNA or dG tracts were added to the
cloning vector pBR322 at its Pst I site according to the conditions
described[11].

Bacterial transformation. The hybridized recombinant plasmids[12] were used to transform E. coli χ1776 (under P2-EK2
conditions as required under earlier National Institutes of Health
Guidelines) as described[12-13]. About 50 recombinants were
obtained per ng of cDNA. Bacteria were selected on Luria agar
plates containing 16 μg of tetracycline per ml[12], and all
tetracycline-resistant bacteria were picked and ordered on gridded
8.5-cm Luria agar plates containing 15 μg of tetracycline per ml.

Screening of recombinant bacteria. Recombinant bacteria were
felt-lifted and transferred to replica plates (containing 8.3-cm
Whatman 541 paper on the surface of Luria agar plates supplemented
with tetracycline at 15 μg/ml). Filters were processed for
hybridization essentially as described by Sippel et al.[14].
Between 30 and 50 filters (2100-3500 colonies) were prehybridized
for 16 hr at 37°C in a mixture of 4 x standard saline citrate
(NaCl/Cit), 44% (vol/vol) formamide, 0.5% Na-DodSO$_4$, and Denhardt's solution[15] containing 200 μg of denatured E. coli DNA
per ml. Hybridization with labeled probes was carried out in the
same buffer using 4-300 x 10^6 cpm of probe for 30 hr. Human
μ-specific probes were prepared by 5' ^{32}P end-labeled partially
hydrolyzed mRNA[16], the synthesis of ^{32}P-labeled single-
stranded (ss) cDNA[17], and nick translation of ds cDNA[18]
inserts rescued by plasmid digestion with Pst I (Bethesda Research
Laboratories, Rockville, Md) and isolated from 5% polyacrylamide
gels[19] by the method of Maxam and Gilbert[3]. After a wash
with prehybridization buffer for 16 hr at 37°C, the filters were
washed batchwise with 6x, 2x, and 0.5x NaCl/Cit, each containing
0.5% NaDodSO$_4$, for 2 hr each at 23°C. The filters were autoradiographed with XRP film for 2-5 days at -20°C. Recombinant
plasmids were isolated from 1-liter stationary phase cultures that
were treated with chloramphenicol (44 μg/ml) for 6 hr prior to
standard CsCl/ethidium bromide banding[20].

Hybrid selection translation. About 50 μg of each recombinant plasmid that was positive to μ probes was covalently linked
to 1-cm discs of diazobenzylmethoxy paper as described[21].
These filters were used with 0.5-1.2 mg of polyadenylated RNA isolated from GM607 cells for selective batchwise hybridization[21]
to their complementary mRNAs. The mRNAs eluted from individual
filters were coprecipitated with 10 μg of yeast tRNA and translated in vitro[8] in the presence of [^{35}S] methionine. The
labeled polypeptides were resolved on 11% reducing NaDodSO$_4$/poly-
acrylamide gels, embedded, and autoradiographed[4].

DNA sequence determination was carried out[3] by G, G+A, T+C,
C, and A>C reactions with 5'-end-labeled Pst I-rescued inserts
that were secondarily cleaved with Ava II, Hha I, Hpa II, or Hind I
and separated on and eluted from gels[19, 3] prior to
base-specific chemical cleavage.

Fusion. We fused 10^7 GM1500-6TG-A12 cells with Ficoll purified lymphocytes derived from 10 ml of peripheral blood of the SSPE
patient in the presence of polyethylene glycol 1000 according to
established procedures. The fused cells were then distributed into
23 wells of a Linbro FB-16-24 TC plate in the presence of HAT selective medium. Hybrids were detected in 20 of the 24 wells. The
hybrids were then propagated in HAT medium and cloned by limiting
dilution. Each independent clone derived from an independent well
was tested for the expression of human immunoglobulin chains and
for the ability to immunoprecipitate measles virus proteins[22].

Immunoprecipitation of secreted human immunoglobulin chains.
Immunoprecipitation and 10% SDS-polyacrylamide gel electrophoresis
of secreted human immunoglobulin chains produced by human hybridomas was carried out as described[4, 22]. Hybridoma cultures were
labeled with 100 μCi ^{3}H-leucine (70 Ci per mmol) per ml for 12
hr. The human immunoglobulin chains were immunoprecipitated with
rabbit anti-human heavy chain antigen using established procedures[4, 22], then separated by 10% SDS-polyacrylamide gel electrophoresis as described elsewhere[4, 22].

Analysis of monoclonal antibodies against measles virus. The
procedures used in immunoprecipitation, similar to those described
elsewhere, were as follows: lysates of virus-infected CV1 cells
labeled with ^{35}S-methionine were used as antigen. Aliquots (25
μl) were mixed with 100 μl concentrated culture fluid and incubated at 37°C for 90 min then at 4°C for 4 hr. Rabbit total
anti-human antibody (25 μl) was then added and the incubation
period repeated. Precipitated polypeptides were collected by cen-
trifugation in an Eppendorf centrifuge for 20 min at 10,000 rpm.
The visible pellet was resuspended and washed three times. After
the final washing, the pellet was suspended in lysis buffer, boiled
for 3 min and electrophoresed on a 10% SDS-polyacrylamide gel in

conditions described elsewhere[23]. After fluorography, dried
gels were exposed to Cronex X-ray film.

RESULTS AND DISCUSSION

Expression of human immunoglobulin chains in mouse x human
hybridomas. As shown in Table 1, mouse x human hybridomas produce
more human heavy chains than the human parental cells. For
example, hybrid DSK-13-ZAS-C8 produces approximately 30 times more
human α chains than the human GM1056 parent. Hybrid ESK-12-1DZ-
63 produces 30 times more α chains than the human GM923 parent.
We had previously mapped the human heavy chain gene cluster to
human chromosome 14[4]. Since the hybridomas produce more human
immunoglobulin chains than the human parental cells, we decided to
attempt to purify the human immunoglobulin mRNAs from the hybrid
cells. Labeled polypeptides synthesized in vitro from partially
purified human immunoglobulin mRNAs following two rounds of oligo
(dT) cellulose chromatography and neutral 5-25% sucrose gradient
fractionation were separated by SDS polyacrylamide
electrophoresis[1]. The human poly (A)$^+$ mRNA from a somatic
cell hybrid αD5-DH11-BC11 was primed with oligo (dT) and reverse
transcribed with avian myeloblastosis virus reverse transcriptase.
The synthesis of the second strand was carried out as described[24]. The synthesized cDNAs were trimmed to blunt ends by
S1 nuclease digestion, tailed with polymeric tracts of dC and
cloned in the plasmid pBR322 at the Pst I site. To assess the
nature of the cloned inserts we have used the hybrid selection
positive translation method[2]. As shown in Figure 1, four
cloned cDNAs were capable to rescue μ specific mRNA. To prove
conclusively that the cloned inserts were μ specific, we have
sequenced the DNA of the cloned by the Maxam and Gilbert

Figure 1. Autoradiogram of [^{35}S]methionine-labeled proteins
synthesized in vitro with mRNAs hybrid-selected from total human B
cell mRNA by recombinant plasmid cDNAs. GM607 polyadenylated RNA
(1 mg) was hybridized with H and L recombinant plasmids linked to
diazobenzylmethoxy-paper and the products of in vitro translation
directed by eluted mRNA were resolved on composite NaDodSO₄/poly-
acrylamide gels. Lanes: 7 and 29, endogenously labeled products
of the reticulocyte lysate; 1 and 24, background translation products of mRNAs eluted from vector-linked filters (pBR322). The
other lanes show the translation products of mRNAs eluted from 28
independent recombinant filters: 6, pTD-Hκ-(607:1-31); 12 and
30, pTD-Hκ-(607:8-14), plasmids that selectively hybridize human
κ mRNA; 9, pTD-Hμ-(αD5:11-10); 10, pTD-Hμ-(αD5:11-16);
23, pTD-Hμ-(607:6-9) and 25, pTD-Hμ-(607:7-12) all selectively
hybridize human μ mRNA. M$_r$ x 10^{-3} of standard proteins are
shown. μ, μ', κ, and κ', positions of in vivo and in vitro
synthesized polypeptides.

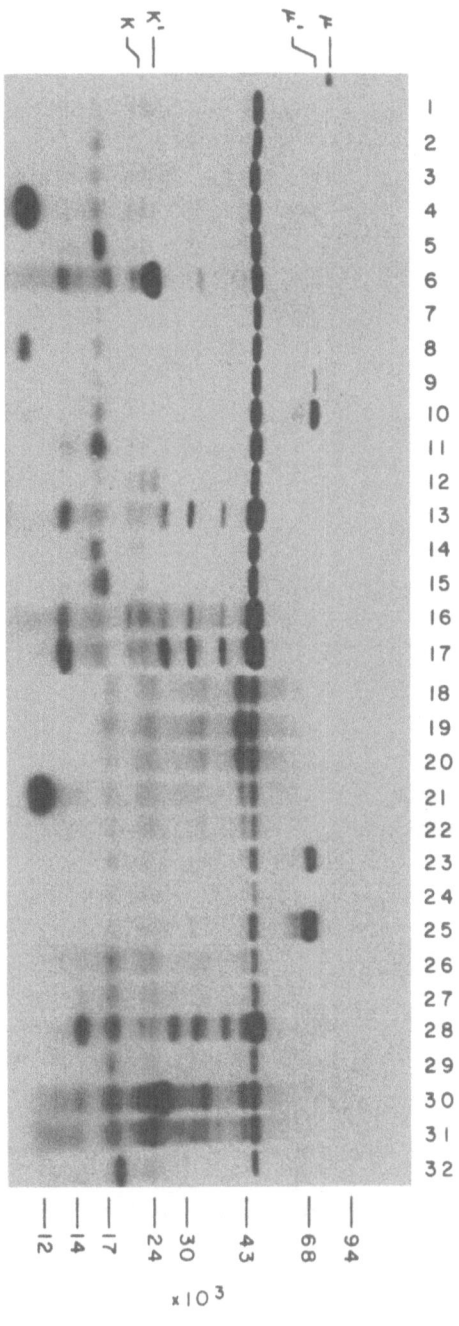

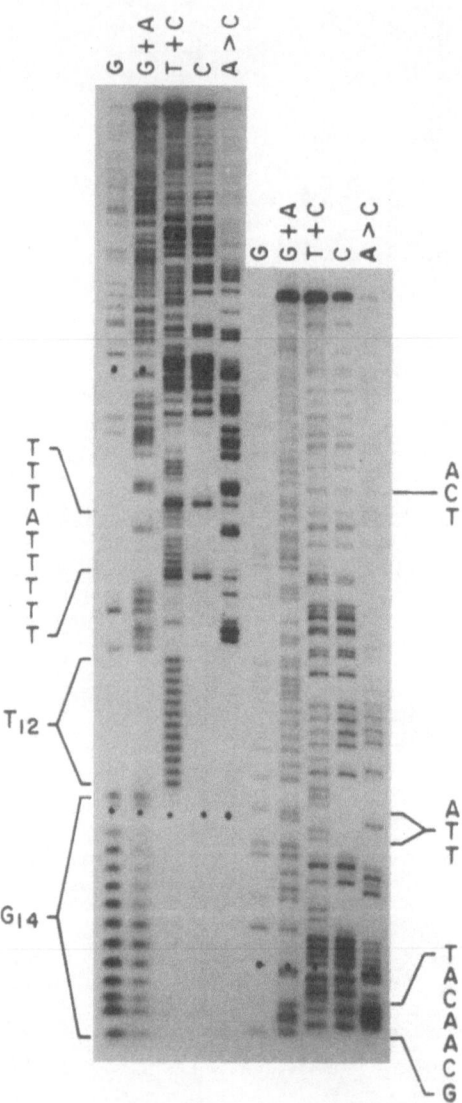

Figure 2. DNA sequencing gel of a μcDNA according to the Maxam and Gilbert method.

```
Residue               421              430                    440
Amino Acid                GlyGluArgPheThrCysThrValThrHisThrAspLeuProSerProLeuLysGlnThr
       5'            GG(31)CCGGGGAGACGTTCACGTGCACCGTGACCCACACAGACCTGCCCTCGCCACTGAAGCAGACC
       3'  ACGTCC(31)GGCCCCTCTCCAAGTGCACGTGGCACTCGGTGTGTCTGGACGGGAGCGGTGACTTCGTCTGG

                          450                    460
                IleSerArgProLysGlyValAlaLeuHisArgProAspValTyrLeuLeuProProAlaArgGluGlnLeu
       5'       ATCTCCCGGCCCAAGGGCGGTGGCCCTGCACAGGCCCGATGTCTACCTGCTGCCACCAGCCCGGGAGCAGCTG
       3'       TAGAGGGCCGGGTTCCCCCACCGGGACGTGTCCGGGCTACAGATGGACGACGGTGGTCGGGCCCTCGTCGAC

                    470                    480
                AsnLeuArgGluSerAlaThrIleThrCysLeuValThrGlyPheSerProAlaAspValPheValGlnTrp
       5'       AACCTGCGGGAGTCGGCCACCATCACGTGCCTGGTGACGGGCTTCTCTCCCGCGGACGTCTTCGTGCAGTGG
       3'       TTGGACGCCCTCAGCCGGTGGTAGTGCACGGACCACTGCCCGAAGAGAGGGCGCCTGCAGAAGCACGTCACC

                    490                    500                         509
                MetGlnArgGlyGlnProLeuSerProGluLysTyrValThrSerAlaProMetProGluPro
       5'       ATGCAGCGGGGGCAGCCCTTGTCCCCGGAGAAGTATGTGACCAGCGCCCCTATGCCGGAACCC-
       3'       TACGTCGCCCCCGTCGGGAACAGGGGCCTCTTCATACACTGGTCGCGGGGATACGGCCTTGGG-

                560                    570                576
                ThrLeuTyrAsnValSerLeuValMetSerAspThrAlaGlyThrCysTyr *
       5'       ACCCTGTACAACGTGTCCCTGGTCATGTCAGACACAGCTGGCACCTGCTACTGACCCTGCTGGCCTGCCCAC
       3'       TGGGACATGTTGCACAGGGACCAGTACAGTCTGTGTCGACCGTGGACGATGACTGGGACGACCGGACGGGTG
                                                                    ‾‾‾‾‾‾‾‾‾‾‾‾

                                                 ‡
       5'       AGGCTCGGGCGGCTGGCCGCTCTGTGTGTGCATGCAAACTAACCGTGTCAACGGGGTCGAGATGTTGCATCT
       3'       TCCGAGCCCGCCGACCGGCGAGACACACACGTACGTTTGATTGGCACAGTTGCCCCAGCTCTACAACGTAGA
                 ‾‾‾‾‾     ‾‾‾‾‾‾‾‾‾‾‾‾‾‾‾‾‾‾‾‾                   ‾‾‾‾‾‾‾‾‾‾‾‾‾‾‾‾‾‾

                                          †
       5'       TATAAAATTAGAAATAAAAAGATCCATTCA(12)C(26)CTGCA
       3'       ATATTTTAATCTTTATTTTTCTAGGTAAGT(12)G(26)G
                 ‾‾‾‾‾‾‾‾‾‾‾‾‾‾‾‾
```

Figure 3. Partial nucleotide sequence of human μ[pTD-
Hμ(αD5:11-16)]-cDNA insert rescued from a Pst I digestion of
the recombinant plasmid. Residue refers to amino acid residue from
NH₂ terminus of human OU. The nucleotide sequence between residues 510 and 559 is pending. *, Termination codon UGA; ‡, termination codon UAA; †, beginning of poly (A) tail. Sequences under-
lined are homologous sequences observed in the mouse μ untrans-
lated sequence.

method[3] (Fig. 2) and found that their nucleotide sequence match the published amino acid sequence of the human immunoglobulin μ constant region (Fig. 3)[1]. These results indicate that it is feasible to clone human immunoglobulin cDNAs by using mRNAs derived from hybridomas. Such cDNAs can be used as probes to clone genomic immunoglobulin DNA sequences and to study the structure and organization of the human immunoglobulin genes in patients with genetic immunodeficiencies and with autoimmune diseases.

We have also used a human B cell line derived from a patient with multiple myeloma (GM1500) (Table 1) to select mutants deficient in hypoxanthine phosphoribosyltransferase. These mutants were then used to produce human x human hybridomas. At first we fused the HPRT deficient GM1500 cells with peripheral lymphocytes derived from a patient with subacute sclerosing panencephalitis (SSPE), a disease caused by measles virus infection of the central nervous system. The patient was a 19 years old female who had typical histological lesions of the brain. Serum from this patient, diluted 1:10[6], bound in radioimmunoassay (RIA) with measles virus infected target cells. The fused cultures were distributed into a Linbro FB-16-24TC plate in the presence of HAT selective medium[24]. Hybrids grew in 20 of the 24 wells. They were propagated in HAT medium and cloned by limiting dilution. Each tested clone derived from an independent well. As shown in Figure 4, the culture fluids of two hybridoma clones (D3 and C5) reacted specifically with the virus nucleocapsid polypeptide (NP) and varying amounts of its cleavage fragments[22]. NP is extremely sensitive to proteolytic cleavage, thus the long incubation periods used in the immunoprecipitation procedure could be partly responsible for the high level of cleavage of the polypeptide. The specificity of the antibodies for NP was confirmed by the demonstration that subclones of hybrid clone D3 could also precipitate this polypeptide (Fig. 5b).

Figure 4. Polyacrylamide gel electrophoretic analysis of the measles virus polypeptides precipitated by the various monoclonal antibodies. a, Virus polypeptides precipitated by convalescent serum of a patient with atypical measles were used as markers. The polypeptides are as follows: H, the virus haemagglutinin; P, a polypeptide associated with the internal nucleocapsid structure; NP, the major structural polypeptide of the nucleocapsid; F, the polypeptide responsible for cell fusion and haemolytic activities; M, the non-glycosylated membrane polypeptide. Bands 1, 2 and 3 are not unique virus polypeptides, but represent proteolytic cleavage fragments of the NP polypeptide[8]. b, Virus polypeptides precipitated by antibody D3. c, Polypeptides precipitated by antibody C5. d, Culture fluid from the human GM 1500 6TG-A12 cell line. Culture fluids of b, c and d were concentrated 20-fold by freeze-drying before use.

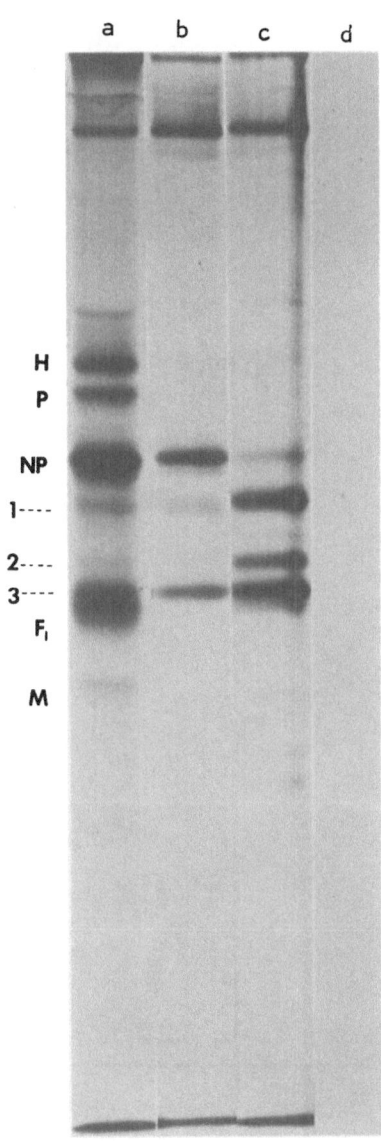

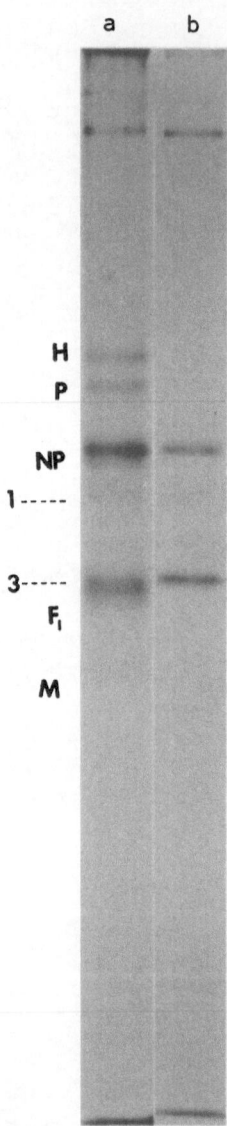

Figure 5. Lane a represents the virus polypeptides precipitated by
the same atypical measles serum as in Fig 2a. b, Polyacrylamide
gel electrophoretic analysis of the measles virus polypeptides
precipitated by culture fluid from subclone D3M2. The conditions
of the experiments were as described in the Materials and Methods.

These results indicate that it is possible to obtain human B cell hybrids which continuously secrete human antibodies, particularly antibodies specific for a human pathogenic virus. The availability of a human continuous B cell line with appropriate drug resistance markers represents a breakthrough in work towards the possible application of this technology to human immunotherapy. In addition, lymphocytes from patients with human autoimmune diseases such as myasthenia gravis and Graves' disease could be fused with human B cell lines to produce hybridomas that secrete the autoantibodies responsible for the disease. Once such monoclonal autoantibodies are available, the production of anti-autoantibodies could provide a cure for patients with these autoimmune diseases.

ACKNOWLEDGEMENT

This work was supported by NIH grants CA10815 and CA23568 and by grant 1-522 from the National Foundation - March of Dimes. We thank Mrs. Jan Erickson for technical assistance.

REFERENCES

1. Dolby, T.W., DeVuono J. and Croce, C.M. 1980. Proc. Natl. Acad. Sci. USA 77:6027-6031.

2. Ricciardi, R.P., Miller, J.S. and Roberts, B.E. 1979. Proc. Natl. Acad. Sci. USA 76:4923-4931.

3. Maxam, A. and Gilbert, W. 1980. Methods Enzymol. 65:499-560.

4. Croce, C.M., Shander, M., Martinis, J., Cicurel, L., D'Ancona, G.G., Dolby, T.W. and Koprowski, H. 1979. Proc. Natl. Acad. Sci. USA 76:3416-3419.

5. Kohler, G. and Milstein, G. 1975. Nature (London) 256:495-497.

6. Stephens, R.E., Pan, C., Ajiro, K., Dolby, T.W. and Borun, T.W. 1977. J. Biol. Chem. 252:166-172.

7. Marcu, K.B., Valbuena, O. and Perry, R.P. 1978. Biochemistry 17:1723-1733.

8. Pelham, H. and Jackson, R.J. 1976. Eur. J. Biochem. 67:247-256.

9. Wickens, M.P., Buell, G.N. and Schimke, R.T. 1978. J. Biol. Chem. 253:2483-2495.

10. Schenk, T.E., Rhodes, C., Rigby, P.W.J. and Berg, P. 1975. Proc. Natl. Acad. Sci. USA 72:984-993.

11. Roychoudhury, R., Jay, E. and Wu, R. 1976. Nucleic Acids Res. 3:101-116.

12. Villa-Komaroff, L., Efstratiadis, A., Broome, S., Lomedico, P., Tizard, R., Naber, S.P., Chick, W.L. and Gilbert, W. 1978. Proc. Natl. Acad. Sci. USA 75:3727-3731.

13. Enea, V., Vovis, G.F. and Zinder, N.D. 1975. J. Mol. Biol. 96:495-509.

14. Sippel, A.Z., Land, H., Lindenmaier, W., Nguyon-Huu, M., Wurtz, J., Timmis, K.N., Giescoke, K. and Schultz, G. 1978. Nucleic Acids Res. 5:3275-3294.

15. Denhardt, D. 1966. Biochem. Biophys. Res. Commun. 23:641-646.

16. Humpheries, P., Old, R., Coggins, L.W., McShane, T., Watson, C. and Paul, J. 1978. Nucleic Acids Res. 5:905-924.

17. Meyers, J.C., Spiegelman, S. and Kacian, D.L. 1977. Proc. Natl. Acad. Sci. USA 74:2840-2843.

18. Rigby, P.W.J., Dieckmann, M., Rhodes, C. and Berg, P. 1977. J. Mol. Biol. 113:237-251.

19. Maniatis, T., Jeffrey, A. and Van de Sande, A. 1975. Biol. Chemistry 14:3787-3794.

20. Kuperstock, Y.M. and Helsinki, D.R. 1973. Biochem. Biophys. Res. Commun. 54:1451-1459.

21. Goldberg, M.L., Liftan, R.P., Stark, G.R. and Williams, J.G. 1979. Methods Enzymol. 68:206-220.

22. Croce, C.M., Linnenbach, A., Hall, W., Steplewski, Z. and Koprowski, H. 1980. Nature 288:488-489.

23. Hall, W.W., Lamb, R.A. and Choppin, P.W. 1979. Proc. Natl. Acad. Sci. USA 76:2047-2051.

24. Littlefield, J.W. 1964. Science 145:709-710.

EXPRESSION OF LIVER MONO-OXYGENASE FUNCTIONS INDUCED BY XENOBIOTICS

M.C. Lechner, C.M. Sinogas, M.T. Freire, and J. Bràz

Instituto Gulbenkian de Ciência

Ap. 14 - 2781 Oeiras, Codex - Portugal

INTRODUCTION

Living cells possess accurately programed mechanisms for regulating the relative amounts of different proteins synthesized. Regulation of gene expression in somatic cells is complex, and is not yet completely understood, affording one of the most interesting and important challenges in contemporary biochemical research. In the highly differentiated eucaryotic cells of vertebrates, a large fraction of the genome is permanently repressed. Only a relatively small fraction of the total genome can be induced or depressed reversibly. These cells do in fact have the capacity for induction of some enzymes by their substrates, although their responses to inducing agents tend to be slow and less dramatic than in procaryotes.

Enzyme induction by substrates is specially evident in the liver of mammals. Actually the liver, the major nutrient-distributing centre for the whole organism, requires induction and repression mechanisms to adapt to the variations of the nutritional intake of the animal. In addition, hepatic cells have a highly developed capacity for enzyme induction and positive regulation of the protein synthesis by hormones. Moreover, most of the detoxification processes in vertebrates, take place in the liver, where lipossoluble foreign compounds are, in a first step bound to the endoplasmic reticulum membranes and subsequently solubilized through hydroxylation and oxidation reactions, catalized by the mixed-function-oxydases system (1). This system consists of an electron transport chain in which a flavoprotein, the NADPH dependent cytochrome c reductase, reduces a specialized microsomal cytochrome, the cytochrome P_{450} (2). Both the flavoprotein and cytochrome P_{450} are inducible by numerous

69

endogenous and exogenous agents (3). Newborn animals miss these
enzymes, which develop after birth (4,5). Exposure to environment
chemicals, drugs, food additives and steroids- can elicit a specific
response of the hepatic endoplasmic reticulum in experimental animals
as well as in man, with liver hypertrophy and induction of the cytochrome P_{450} dependent biotransformation systems. This constitutes
a typical case of enzymatic adaptation in animals.

The cytochrome P_{450}'s are integral membrane proteins of the
endoplasmic reticulum. Evidence now indicates that multiple forms
of cytochrome P_{450}'s exist in the liver microsomes, which are inducible to different extents by different agents (6). The mixed con-
stituency of types of cytochrome P_{450} recognized explains, at least
in part the fact that these key components of the liver biotransformation system are responsible for the metabolism of so many different
substrates as drugs, mutagens, steroids, etc. The sequences of P_{450}
dependent metabolic pathways are not fundamental to maintenance of
life. Fundamental metabolic sequences such as glycolysis, and protein
biosynthesis should be present in all cells, but, the cytochrome P_{450}
dependent processes are related to differentiated functions. They
are not primitive, being distributed only in cells having specialized
functions.

INDUCTION BY POLYCYCLIC HYDROCARBONS AND THE Ah LOCUS

3-methylcholantrene (MC) and other polycyclic aromatic compounds
are well known inducers of the microsomal mono-oxygenase systems
such as aryl-hydrocarbon-hydroxylases (AHH). Induction of AHH activity is dependent of a specific form of cytochrome P_{450} known as
P_1-_{450} or P_{448} (7).

A single genetic site has been postulated by NEBERT and co-
-workers, the Ah-locus, to controle the induction of three enzymes
(8). According to NEBERT's hypothesis, polycylic aromatic hydrocarbons bind to a specific receptor in the cytosol, and the resulting complex is translocated into the cell nucleus, similarly to the
mechanism of action of steroid hormones (9). The responsiveness to
aromatic hydrocarbons was designated the Ah locus, numerous studies
indicating that an important product of the Ah regulatory locus in
mice is the cytosolic receptor capable of binding to certain polycyclic aromatic inducers. This hypothesis is supported by the findings of a stereospecific high-affinity binding of 2,3,7,8-tetra-
chlorodibenzo-p-dioxin (TCDD) which acts through the same genetic
site as other polycyclic hydrocarbons, in different strains of mice
which have the Ah^b allele, and correlating well to the AHH inducibility (10). The TCDD-receptor complex was found to interact with
DNA, this interaction resulting in a specific induction of the
cytochrome P_{450} dependent AHH (11). Activation of structural genes,
leads to increases in enzymes which metabolize the inducer as well

as other polycyclic aromatic compounds. The induction of at least
20 mono-oxygenase activities is closely associated with the Ah locus.
Other inducible enzymes that are not mono-oxygenases appear to require
the same cytosolic receptor protein, including microsomal UDP-glucu-
ronyltransferase, cytosolic NADPH menadione oxidoreductase and ornithine decarboxylase. This class of inducers binding to a cytosolic
receptor evoke a pleiotropic response (12,9).

The AHH assay constitutes a sensitive test for aromatic hydrocarbon responsiveness to polycyclic hydrocarbon inducers. AHH induction can therefore be used as an indicator of phenotype at Ah locus.
Using this criterium, responsive and non-responsive inbred mouse
strains have been found. Non-responsive strains lack the gene product
of the Ah locus, the cytosolic receptor molecule (13,14).

The Ah locus plays a major role in the susceptibility of the
animals to MC and benzpyrene (BP) tumorigenesis, although other genes
seem to be responsible for the increased susceptibility to these
induced tumors (15,16). MC and BP are metabolized to carcinogenic
intermediates, as the 7,8-diol-9,10-epoxide of BP, predominantly
by P_{1-450} mediated mono-oxygenase (17), the diol being previously
formed under epoxide hydrase catalysis. The relative content of
P_{1-450} compared with other forms of P_{450} may be specially large in
tissues like skin, where the radio is 10:1 and lung, 50:1 but never
reaches even a ratio of 1:1 in the liver. Mouse lung is known to be
susceptible to MC tumorigenesis and a statistically significant
correlation between lung tumors produced by intratracheal MC and
the Ah allele has been found (18).

There is now immunological evidence that in rat liver, a minimum
of six different forms of P_{450} exist which have been isolated (19,
20,21). But the statistical analysis of partially separated mono-
-oxygenases activities, leads to the conclusion that a minimum of
20 forms of P_{450}, structural and inducible, must co-exist in mouse
liver (22).

Distinct cytosolic receptor species binding to differente mono-
-oxygenases inducers have been demonstrated to be present in rat
liver cytosol, by isoelectric focusing in PAGE (10). The binding
is competed by TCDD, MC and β-naftoflavone, however phenobarbital
(PB) does not compete to the same binding sites, and as yet there
is no report of PB receptor protein present in hepatocytes analogous
to the steroid receptors.

INDUCTION BY PHENOBARBITAL

PB is an inducer of the liver mixed-function-oxidases, which
has been widely used as the prototype of another group of liver
microsomal enzyme inducers (23).

When given to the experimental animals it strongly stimulates the production of increased amounts of the mixed-function-oxidases (24,25). This enzymes induction is concomitant with the hyperthrophy of the organ due to a marked proliferation of the E.R. membranes, providing a model which has been largely used for the study of membrane biogenesis and cell development.

PB affects the protein pattern of the liver cell by selectively inducing the synthesis of some endoplasmic reticulum components (26), without affecting the amino-acid incorporation into the proteins of other sub-cellular compartments (27). Chronic PB administration brings about a progressive rise of the induced proteins which attain a high steady sate at 72 hours. Cytochrome P_{450} the terminal catalytic component of the liver microsomal mono-oxygenase system is preferentially induced, attainning up to 200% of the normal value in adult rat liver E.R. (28). However, the inductive effect of PB is not confined to the catalytic components of the mixed-function-oxidases, as this chemical agent induces hypertrophy of the liver cell with intense proliferation of the E.R. associated to an important increase of the total microsomal proteins. The induction by PB is the result of an increase in the protein synthesizing activity of the liver microsomes, which are the main site for protein synthesis in the liver cell. A single dose of PB elevates more than twofold the synthesis of nascent peptides on membrane-bound polysomes, 3-4 hr. after treatment with the inducer (29), while protein synthesis by free polysomes does not present significative changes under PB action (30).

The liver hypertrophy and enzyme induction by PB, depend on important changes in the metabolism of macromolecules. An increase in the RNA content of the cells is observed, preceeding the enhancement in protein biosynthesis (31). The half-life of RNA in the liver is increased by pretreatment with this inducer (32) and inversely related to the changes of microsomal alkaline ribonuclease produced during the stimulation of drug metabolism by PB (28). These phenomena are associated to important reorganization of the E.R. membranes (33). However it is not clear whether the alterations in protein synthesis are primarily due to an enhancement in the stability of mRNA's, to an increase in the rate of translation of pre-existing mRNA's or to an increase in the synthesis of required mRNA's.

We have been interested in the study of the course of biochemical events concerning gene expression and regulation taking place during the onset of induction by PB, in order to search for the primary effects produced by this agent, which trigger the complex modifications accounting for the modulation of protein synthesis and consequent enzyme induction and E.R. membranes proliferation, brought about by this xenobiotic.

Our studies have indicated that the synthesis of membrane proteins induced by PB is associated to a marked increase in poly(A)$^+$ mRNA in the liver microsomes (34).

RNA labelling kinetics *in vivo* studied by administration of [^{14}C] orotic acid, as well as determination of template activities of the microsomal membranes and their RNA's (35), have shown that an accumulation of active mRNA is produced shortly after PB administration at this sub-cellular level. However, when RNA synthesis was studied by [^{14}C] orotic acid incorporation into nuclear RNA, we failed to prove an enhancement in the synthesis. The amount of labelled precursor incorporated into total nuclear RNA was lower in the livers of PB induced animals than in controls, the specific radioactivities found being 13-27% lower, over an incorporation period of 0,5 - 2hrs. Total label incorporated per cell, calculated on the basis of nuclear DNA, was 37,21 and 1% lower, at 0.5, 1.0 and 2.0 hrs respectively (36).

Newly synthesized RNA, primary transcripts, as well as intermediary processing pre-mRNA do not accumulate inside nuclei (37), however our results indicate that the trigger for the stimulatory effect of PB on protein biosynthesis must not be an enhancement in RNA transcription. Consistent with this are the absence of stimulation of nuclear RNA polymerases I and III as well as of polymerase II, over the period between 3 hours and 4 days after PB administration (38), and the fact that any discernible increase in [^{32}P] incorporation into nuclear, nucleolar and nucleoplasmic RNA has been detected during induction by PB (39).

Gene expression in eukaryotic cells is regulated to a large extent at post-transcriptional levels. The existence of precursors to mRNA's containing by far more RNA sequences than the functional mRNA, and particularly the discontinuous form in which the coding sequences are stored at DNA level, preclude the existence of complex mechanisms of processing in which recognition of informative and regulatory sequences in each specific messenger are compulsory (37).

Polyadenylation represents a general service mechanism common to pre-mRNA and mRNA, independent of the processing stage, cellular compartment and repression or translation of the messenger in cytoplasm. For the most part of mRNA's in the liver as in other somatic cells, transport of mRNA from nucleus to the cytoplasm is dependent on polyadenylation (40).

The activity of the poly(A) polymerase has been studied in the isolated nuclei. If the accumulation of poly(A)$^+$ mRNA in the cytoplasmic compartment previously observed were due to an acceleration of intra-nuclear processing of mRNA precursors, an enhancement in

the activity of the poly(A) polymerase in the nuclei from PB induced
rats would be expected to be found. However, our results revealed
that any detectable activation of the intra-nuclear polyadenylation
system is caused by this agent (41).

Messenger RNA exists in animal cells as ribonucleoprotein
complexes in different pools. Informosomes, the mRNP particles in
the cytoplasm of secretory cells exist not only in polyribosomal
free and bound mRNP's, but also as free short-term and long-term
repressed mRNA, absent from the polyribosomal mRNA population. These
non-polyribosomal informosomes can be bound to the E.R. membranes
or free in the cytoplasm.

A kynetic relationship between non polyribosomal "silent"
messenger RNA and polyribosomal mRNA, consistent with a precursor-
-product relationship between the respective mRNP's has been found
in different biological models (42). Actually several different
mRNA's species, coding for specific proteins have been identified
in the post-ribosomal supernatant of a variety of tissues.

The existence of these potentially functional mRNA's stored
in the cytoplasm in a repressed state as mRNP complexes constitutes
an important device for translational control mechanisms, as there
is a reversible equilibrium between polysomes and free mRNP plus
ribosomal sub-units. A pool of poly(A)$^+$ free mRNP particles exists
in rat liver, to which a biological role as a stable precursor of
active polyribosomal membrane bound messenger is ascribed (43).
Post-ribosomal supernatant of normal rat liver contains 15% of the
total poly(A)$^+$ rich RNA present in cytoplasm. A large pool of the
total cytoplasmic ferritin mRNA (44%) is present in the post-
-ribosomal supernatant of normal rat liver and a cytoplasmic control
mechanism has been proven to exist for iron stimulated ferritin
synthesis in the liver (44). Iron treatment causes a dramatic
drecrease in the post-ribosomal ferritin mRNA with a corresponding
increase in the polyribosomal ferritin pool. Also the proportion
of albumin mRNA present in the liver post-ribosomal supernatant
fraction increases dramatically in a short term fast, representing
up to 60% of total cytoplasmic albumin mRNA sequences. Albumin mRNA
is therefore probably stored in the mRNP fraction during the fasting
state as the reduced rate of albumin synthesis in fasting can be
rapidly reversed by feeding or by supplementation of the animal
with amino acids, through a rapid reassembly of polyribosomes (45).

During induction by PB in the liver a two-fold increase in the
rate of *in vivo* cytochrome P$_{450}$ apoprotein synthesis is observed
6 hours after the administration of PB (46) while the lag period
for detecting increased amounts of *in vivo* translatable mRNA in
the total poly(A)$^+$mRNA extracted has been demonstrated to be about
16 hours (47).

We have previously observed that total poly(A)$^+$mRNA isolated
from rat liver, 24 hours after PB administration are more active
in stimulating amino-acids incorporation into several proteins including cytochrome P_{450} inducible apoprotein, when assayed *in vitro*
in a reticulocyte lysate translation system (48). We have previously
observed that poly(A)$^+$RNA in the free cytoplasmic RNP particles
decreases to 68% of the value found in non-induced rat liver, 24h
after a single administration of PB (80mg/kg body wt.) while concomitantly poly(A)$^+$RNA associated to the E.R. membranes rises to
157% of the normal (36). Labelling kynetics studies of the
poly(A)$^+$RNA in the free cytoplasmic RNP particles revealed the
presence of large amounts of *de novo* synthesized messengers, suggesting that at least part of these free informosomes pool consists of
potentially active PB inducible messages, coding for E.R. proteins
(49).

Translation of free cytoplasmic RNP's and their poly(A)$^+$RNA's
in a reticulocyte lysate cell-free-system revealed that the samples
from PB induced animals display a higher and durable *in vitro*
template activity, representing approximately 200% of the normal
value for a 60 min. incubation time (36).

It is known that 3'OH poly(A) tails plays a fundamental role
in the biogenesis of mRNA and its utilization for translation in
eukaryotic cells, the length of these structures being inversely
related to the age of the mRNA molecule (50). We have determined
the relative amount of the poly(A) sequences in the total poly(A)$^+$
RNA from free informosomes, by molecular hybridization tests, using
$|^3H|$ poly U. A value of 7,1% poly(A) has been found for the RNA
from PB free cytoplasmic particles, representing 136% of the value
found in non induced particles, which was 5.24% (36).

Our results demonstrate that the free RNP pool in the liver
is markedly affected by administration of PB and point to a role
of these particles as a stock of potentially active messages,
vehicles for regulation mechanisms displayed during induction by
this agent.

The higher poly(A) percentage found in the informosomes after
induction might determine a stronger affinity to the E.R. membranes,
contributing to the enhancement in protein synthesis found *in vivo*
at this sub-cellular level in the induced livers.

It is known that the direct association of mRNA and membranes
can contribute to the spacial segregation of messengers to be translated on ribosomes bound to endoplasmic reticulum (51,52) and it
has been recently demonstrated that cytochrome P_{450} mRNA in PB
induced rat liver is primarily associated with ribosomes bound to
the E.R. membranes the cytochrome P_{450} apoprotein being exclusively

synthesized (>95%) by polysomes associated with the E.R., and
directly inserted into the membranes (53).

We have searched for a possible precursor-product relationship
between free-cytoplasmic RNP's and bound polysomal mRNA responsible
for the synthesis of induced proteins by performing double-cross-
-experiments between RNP particles and stripped E.R. membranes from
induced and non induced rat livers.

An important increase in the capacity of stripped microsomes
from the induced livers, to bind RNP complexes has been found (164%)
while it has been demonstrated that RNP particles from induced and
non induced livers do not significantly differ in their binding
aptitude. The membranes from the PB induced livers also exhibited
a much higher capacity to bind polysomes which value was 322% of
the normal (36).

The increase in the attachement of mRNA to the membranes may
insure its proximity to the ribosomal binding sites, improving the
translation of many different proteins by facilitating the reutiliza-
tion of the messengers (54). We have studied the time course of the
E.R. membranes capacity to bind RNP complexes, the results demonstrating that PB evokes a very important increase in this binding
capacity at a very early stage of the inductive response. Actually,
two hours after administration of the inducer, a threefold enhancement of the f.c. RNP's binding is observed, while polysomes binding
attains a twofold value at 12 hours (55).

From these experiments we conclude that a primary effect of
PB in the liver would be the movement of stocked mRNA from f.c.
RNP's into bound polysomes.

We have previously shown that after PB, there is a very important and early reduction in the alkaline microsomal RNase, preceding
the rise of all the induction indexes(28) which is apparently related
to a decrease in RNA degradation, demonstrated both *in vitro* and
in vivo (56,57). A prolongation of the half life of total and messenger RNA has been proven to be brought about by the barbiturate
(58).

The level of each component inside a cell is the result of a
dynamic equilibrium between synthesis and catabolism, and many
control mechanisms can operate at a post-transcriptional level,
including mRNA processing and exportation, as well as mRNA lifespan
and degradation. Actually there is ample evidence for varying half
lives of different mRNA within one cell, and mRNA lifespan in
eukaryotes vary between minutes and months most of them being translated many times. Therefore, controled degradation of different
mRNA molecules must constitute an important device in the regulation
of protein synthesis.

Together our results suggest that the overall acceleration in
the rate of protein synthesis induced by PB, may be the consequence
of primary effects at the level of the E.R. membranes, where this
drug is known to accumulate immediately after being absorbed (25).
The observed increase in the capacity of microsomes to bind RNP
complexes may be the result of a reorganization of the membranes
evoked by this xenobiotic. As summarized in the figure 1, we admit
that the presence of PB or its metabolites at the E.R. membranes
may be directly responsible for the enhancement in the binding
capacity of the membranes, by changing their electrostatic properties.

Alternatively the transient enhancement of the phospholipid/
cholesterol ratio produced at the early stages of PB induction (59)
may be postulated as an important factor determining the increase
in the binding capacity of the E.R. membranes to RNP complexes, as
cholesterol is known to be a degranulating agent impairing polysome
binding (60). The improved fluidity of the membrane brought about
by the phospholipids would be an additional factor for the efficiency
of translation by bound polysomes as there is some evidence suggesting that membranes play a role in extending the functional availability of templates and stabilization of mRNA (54).

The increase in the affinity of the E.R. membranes to RNP
complexes may be responsible for shifting the equilibrium between
the different forms of mRNA's in the cytoplasm-latent and active-
towards its entry into bound polysomes. This could originate gene
activation mechanisms arising by feed-back control, as we have
recently observed a progressive increase of the poly(ADP-P) polymerase activity, in rat liver nuclei, following PB administration
(61,62) an enzymatic system which is responsible for the covalent
modification of histones (63,64) and other nuclear proteins (65)
to which a role in the regulation of chromatin function and in the
metabolism of proteins and nucleic acids has been ascribed (66).

The response to PB is complex, being probably the result of
more than one action produced by this agent and contributing to
the overall acceleration of protein synthesis. We admit that PB
or its metabolites may concomitantly act as direct effectors on
RNA and pyrimidine nucleotides metabolism. The inhibition of the
microsomal alkaline RNase previously observed (28) may be the
result of a competitive inhibition by the barbiturate derivatives
which mimic pyrimidine nucleotides, the sites for the enzyme recognition in the RNA molecules.

These facts could in a large extent explain how.PB produces
hypertrophy in the liver. However, the exact mechanism for the induction of each specific protein affected may vary, and the study
of transcriptional and translational events concerning each mRNA
species has to be performed.

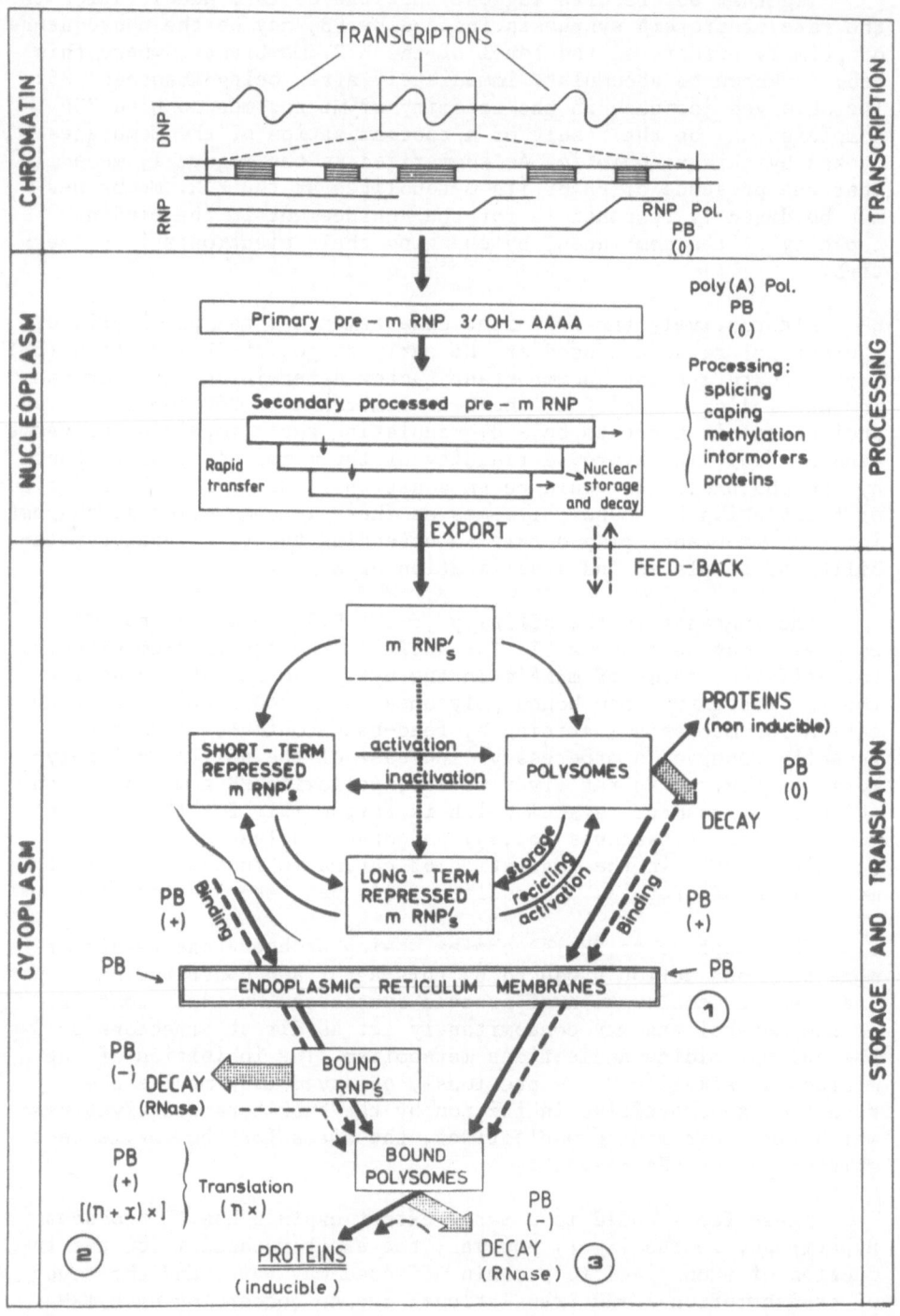

The use of specific cDNA probes for measuring the number of copies of each mRNA in the different sub-cellular compartments and at different stages of the inductive response will decisively contribute to a better understanding of the molecular basis of induction of each particular protein affected by this agent.

It is known that, in rat liver, PB stimulates the production of a particular form of cytochrome P_{450}, the cytochrome $P_{450}b$ (20). Translation of cytochrome $P_{450}b$ apoprotein has been achieved in a reticulocyte lysate cell-free-system, and it has been observed that PB enhances the translation product of this specific message when total poly(A)$^+$mRNA (43) or polysomal poly(A)$^+$mRNA (47) are used as templates *in vitro*. Cytochrome $P_{450}b$ is therefore a good model for the study of the molecular mechanisms induced by PB. Actually, it has been recently demonstrated that this protein represents about 2% of total cytochrome P_{450} in control animals while after PB this value rises to 57% (67). Several groups are presently engaged in obtaining homogeneous preparations of messenger RNA for this inducible protein. Recently Fujii-Kuriyama et als.(68), attempted to clone the cDNA sequence of this mRNA. The obtention of a recombinant plasmid will not only provide an important mean for further work on the molecular mechanism of the inductive response to xenobiotics, but also to describe the structure and organization of this mRNA and respective genes.

The use of cultured hepatocytes retaining their ability to respond to PB, as achieved by Michalopoulos et al. (69), with primary cultures of hepatic cells from the rat, offers an interesting system, for its relative simplicity, to the study of the mechanisms regulating mono-oxygenase induction and genetic control of the different forms of cytochrome P_{450}.

Fig. 1. *Diagram of the flow of mRNA formation and expression of mRNP's in the liver according to the general scheme outlined by Scherrer et al. (37,42). The points where significative changes have been observed under PB action are indicated in the figure* :-1)-a very important increase in the RNP binding capacity of E.R. membranes occurs 2 hrs after PB administration (55), indicating that the mobilization of stored mRNA's, may be the trigger for the complex modifications in gene expression and modulation of protein synthesis induced by this agent. -2)- increase in the translation efficiency of the mRNP's associated to a higher percentage of poly (A) in its RNA (36,48) -3)- decrease of microsomal alkaline RNase activity (28), probably related to the prolongation of mRNA and polysomes half-life.
Detailed explanations are given in the text.

REFERENCES

1) WILLIAMS, R.T. (1959) Detoxification mechanisms, in the Meta-
    bolism and Detoxification of Drugs, Toxic Substances and other
    Organic Compounds, 2nd. ed. John Wiley & Sons, New York.
2) HOEVEN van der, T.A., COON,M.J. (1974) Preparation and Pro-
    perties of Partially Purified Nicotinamide Adenine Dinucleotide
    Phosphate - Cytochrome P_{450} Reductase from Rabbit Liver Micro-
    somes. J. Biol. Chem., 249, 6302.
3) MANNERING, G.J. (1971). Properties of Cytochrome P_{450} as
    affected by environmental factors: qualitative changes due to
    administration of Polycyclic Hydrocarbons. Metabolism, 20, 228.
4) FOUTS, J.R., ADAMSON, R.H. (1959). Drug Metabolism in the
    Newborn Rabbit. Science, 129, 897.
5) DALLNER, G., SIEKEVITZ, P. and PALADE, G.E. (1966) Biogenesis
    of Endoplasmic Reticulum Membranes II. Synthesis of Constitu-
    tive Microsomal Enzymes in Developing Rat Hepatocyte. J. Cell
    Biol. 30, 97.
6) LU, A.Y.H. (1979). Multiplicity of Liver Drug Metabolizing
    Enzymes. Drug Metabolism Reviews, 10, 187.
7) GUENGERICH, F.P. (1977). Separation and purification of multiple
    forms of microsomal cytochrome P_{450}- Activities of different
    forms of cytochrome P_{450} towards several compounds of environ-
    mental interest. J. Biol. Chem., 252, 3970.
8) NEBERT. D.W., JENSEN, N.M. (1979). The Ah Locus: genetic regu-
    lation of the metabolism of carcinogens, drugs and other envi-
    ronmental chemicals by cytochrome P_{450} mediated mono-oxygenases
    CRC Critical Reviews in Biochemistry, Fasman, G.D. ed., CRC
    Press, Inc. Cleveland, Ohio, 6, 401.
9) KUMAKI, K., JENSEN, N,M. SHIRE, J.G.M., NEBERT, D.W. (1977).
    Genetic Differences in Induction of cytosol reduced-NAD(P):
    Menadione Oxydoreductase and Microsomal Aryl Hydrocarbon
    Hydroxylase in the Mouse. J. Biol. Chem. 252, 157.
10) POLAND, A., GLOVER, E. (1975). Genetic expression of Aryl
    Hydrocarbon Hydroxylase by 2,3,7,8, - Tetrachlorodibenzo-p-
    -dioxin: Evidence for a receptor Mutation in Genetically
    Non-responsive Mice. Mol. Pharmacol, 11, 389.
11) CARLSTEDT-DUKE, J., GILLNER, M., HANSSON, L.-A, TOFTGÅRD, R,
    GUSTAFSSON, S., HÖGBERG, B., GUSTAFSSON, J.-Å.(1980). The
    molecular basis for the induction of aryl hydrocarbon hydro-
    xylase: characteristics of the receptor protein for 2,3,7,8
    -tetra- chlorodibenzo-p-dioxin (TCDD) in Biochemistry, Bio-
    physics and Regulation of Cytochrome P_{450} J.Å. Gustafsson et
    al. eds. Elsevier/North Holland, 147.
12) NEBERT, D.W., ROBINSON, J.R., NIWA., KUMAKI., POLAND, A.P.
    (1975). Genetic expression of aryl hydrocarbon hydroxylase
    activity in the mouse. J.Cell Physiol., 83, 393.
13) POLAND, A., GLOVER, E., KENDE, A.S. (1976). Stereospecific,
    high affinity binding of 2,3,7,8-tetrachlorodibenzo-p-dioxin
    by hepatic cytosol: evidence that the binding species is the

receptor for the induction of aryl hydrocarbon hydroxylase.
J. Biol. Chem. 251, 4936.

14) OKEY, A.B., BONDY, G.T., MASON, M.E., KAHL, G.F., EISEN, H.J.,
GUENTHNER, T.M., NEBERT, D.W. (1979). Regulatory Gene Product
of the Ah Locus. Characterization of the cytosolic inducer-
-receptor complex and evidence for its nuclear translocation.
J. Biol. Chem., 254, 11636.

15) KOURI, R.E. RUDE, T.H., JOGLEKAR, R., DANSETTE, P.M., JERINA,
D.M. ATLAS, S.A., OWENS, I.S. NEBERT, D.W. (1978). 2,3,7,8 -
-Tetrachlorodibenzo-p-dioxin acts as cocarcinogen in causing
3-methylcholanthrene-initiated subcutaneous tumors in mice
genetically "nonresponsive"at ah locus:Cancer Res. 38, 2777.

16) NEBERT, D.W., ATLAS, S.A., GUENTHNER, T.M., KOURI, R.E. (1978).
The Ah locus: genetic regulation of the enzymes which metabolize
polycyclic hydrocarbons and the risk for cancer, in Polycyclic
Hydrocarbons and Cancer: Chemistry, Molecular Biology and
Environment , Ts'o, P.O.P. and Gelboin, H.V., Eds., Academic
Press, New York, 345.

17) KOURI, R.E. (1976) Relationship between levels of aryl hydrocarbon hydroxylase activity and susceptibility to 3-methyl-
cholanthrene and Benz[a]pyrene-induced cancers in inbred
strains of mice, in Polynuclear Aromatic Hydrocarbons:
Chemistry, Metabolism and Cancerigenesis, Freudenthal, R.I.,
and Jones, P.W. Eds., Raven Press, New York, 139.

18) KOURI, R.E. NEBERT, D.W. (1977). Genetic Regulation of susceptibility to polycyclic hydrocarbon-induced tumors in the
mouse, in Origins of Human Cancer, Hiatt, H.H., Watson, J.D.
and Winsten, J.A., eds. Cold Spring Harbor Laboratory, New
York, 811.

19) KAWALEK, J.C., LEVIN, W., RYAN, D., THOMAS, P.E., LU, A.Y.H.
(1975). Purification of liver microsomal cytochrome P_{448} from
3-M.C. treated rabbits. Mol. Pharmacol., 11, 374.

20) RYAN, D.E., THOMAS, P.E. KORZENIOWSKI, D., LEVIN, W. (1979).
Separation and characterization of highly purified forms of
liver microsomal cytochrome P_{450} from rats treated with poly-
chlorinated biphenyls, phenobarbital and 3-Methylcholanthrene.
J. Biol. Chem. 254, 1365.

21) THOMAS, P.E., REIK, L.M. RYAN, D.E., LEVIN, W. (1981). Regulation of three forms of cytochrome P_{450} and epoxide hydratase
in rat liver microsomes. Effects of age, sex and inductions.
J. Biol. Chem., 256, 1044.

22) LANG, M.A., NEBERT, D.W. NEGISHI, M. (1980). Structural gene
products of the Ah complex. Separation of multiple forms of
liver microsomal cytochrome P_{450} and characterization of mRNA
associated with P_1-450 from 3-methylcholanthrene-treated mice.
in Biochemistry, Biophysics and Regulation of Cytochrome P_{450}.
J.A.Gustafsson et al. eds. Elsevier/North Holland. 415.

23) MANNERING, G.J. (1968).Significance of stimulation and inhibition of drug metabolism in pharmacological testing -in
Selected Pharmacological Testing Methods. Ed. M. Burger,

Dekker– New York. 51.

24) ORRENIUS, S., ERNSTER, L. (1964). Phenobarbital induced synthesis of the oxidative demethylating enzyme of rat liver
microsomes. Biochim. Biophys. Res. Commun. 16, 60.

25) ERNSTER, L., ORRENIUS, S. (1965). Substrate induced synthesis
of the hydroxylating enzyme system of liver microsomes. Fed.
Proc. 24, 1190.

26) KATO, R., LOEB, L. GELBOIN, H.V. (1965). Microsome-specific
stimulation by phenobarbital of amino-acid incorporation in
vivo. Biochem. Pharmacol., 14, 1164.

27) ARIAS, I.M., DOYLE, A., SCHIMKE, R.T. (1969). Studies on the
synthesis and Degradation of Protein of the Endoplasmic
Reticulum of Rat Liver. J. Biol. Chem., 244, 3303.

28) LECHNER, M.C., POUSADA, C.R. (1971). A possible role of liver
microsomal alkaline ribonuclease in the stimulation of oxidative drug metabolism by phenobarbital, chlorodane and chloro-
phenothane (DDT). Biochem. Pharmacol., 20, 3021.

29) GLAZER, R.I., SARTORELLI, A.C. (1972). The effect of Pheno-
barbital on the Synthesis of Nascent Protein on Free and
Membrane-Bound Polyribosomes of Normal and Regenerating Liver.
Molec. Pharmacol., 8, 701.

30) McCAULEY, R., COURI, D. (1971). Early effects of phenobarbital
on cytoplasmic RNA in rat liver. Biochim. Biophys. Acta, 238,

31) HOLTZMAN, J., GILLETTE, J.R. (1968). The effect of Phenobarbital
on the turnover of Microsomal Phospholipid in Male and Female
Rats. J. Biol. Chem., 243, 3020.

32) STEELE, W.J., (1970). Phenobarbital induced prolongation of
the half-life of ribosomal-RNA of rat liver. Fed. Proc. Fed.
Am. Societies Exp. Biol., 29, 737.

33) SCHIMKE, R.T. (1973). Control of enzyme levels in mammalian
tissues. Adv. Enzymol., 37, 135.

34) LECHNER, M.C. (1976). Effect of phenobarbital treatment on
poly(A)-rich RNA in rat liver microsomes. I.U.B. Xth. Int.
Congr. Biochem., HAMBURG, 03-6-130.

35) LECHNER, M.C. (1974). Studies of RNA from rat liver endoplasmic
reticulum sub-fractions. Effect of phenobarbital treatment.
Naunyn-Schmiedeberg's Arch. Pharm. Supp. 285, R50.

36) LECHNER, M.C., SINOGAS, C.M. (1980). Changes in gene expression
during liver microsomal enzyme induction by phenobarbital. in
Biochem. Biophys. and Regulation of Cytochrome P_{450}. J.A.
Gustafsson et al. eds. Elsevier/North-Holland, 405.

37) SCHERRER, K., IMAIZUMI-SCHERRER, M.T., REYNAUD, C.A., THERWATH,
A.. (1979). On pre-messenger RNA and Transcriptions A review.
Molec. Biol. Rep., 5, 5.

38) LINDRELL, T.J., ELLINGER, R., WARREN, J.T., SUNDHEIMER, D.,
O'MALLEY, A.F. (1977). The effect of acute and chronic pheno-
barbital treatment on the activity of rat liver DNA dependent
RNA polymerases, Molec. Pharm., 13, 426.

39) KUMAR, A., SATYANARAYANA RAO, R., PADMANABAN, G. (1980). A
comparative study on the early effects of Phenobarbital and

3-Methylcholanthrene on the synthesis and transport of ribonucleic acid in rat liver. Biochem. J., 186, 81.

40) SCHUMM, D.E., WEBB, T.E. (1974). Modified messenger ribonucleic acid release from isolated hepatic nuclei after inhibition of polyadenylate formation. Biochem. J., 139, 191.

41) LECHNER, M.C., SINOGAS, C.M. (1978). Studies on liver poly(A)--rich RNA during microsomal enzyme induction. 12th. FEBS Meeting DRESDEN, July 2-8, 1157, (124).

42) MAUNDRELL, K., MAXWELL, E.S., CIVELLI, O., VINCENT, A., GOLDENBERG, S, BURI, J.F. IMAIZUMI - SCHERRER, M.T., SCHERRER, K. (1979). Messenger RNP complexes in avian erythroblasts: Carriers of post-transcriptional regulation? Molec. Biol. Rep., 5, 43.

43) HEMMINKI, K. (1975). Labbeling kynetics of RNA containing poly (A) in liver subcellular fractions. Molec. and Cell. Biochem., 8, 123.

44) ZÄHRINGER, J., BALIGA, B.S., MUNRO, H.N. (1976). Novel mechanism for translation control in regulation of ferritin synthesis by iron. Proc. Natl. Acad. Sci., 73, 857.

45) YAP, S.H., STRAIR, R.K., SHAFRITZ, D.A. (1978). Effect of a short term fast on the distribution of cytoplasmic albumin messenger ribonucleic acid in rat liver. J. Biol. Chem., 253 4944.

46) BHAT, K.S., PADMANABAN, G. (1978). Cytochrome P_{450}, synthesis in vivo and in a cell-free system from rat liver. FEBS Lett., 89, 337.

47) DUBOIS, R.N., WATERMAN, M.R., (1979). Effect of phenobarbital administration to rats on the level of the in vitro synthesis of cytochrome P_{450} directed by total rat liver RNA. Biochem. Biophys. Res. Commun., 90, 150.

48) LECHNER, M.C., FREIRE, M.T., GRONER, B. (1979). In vitro biosynthesis of liver cytochrome P_{450} mature peptide sub-unit by translation of isolated poly(A)+mRNA from normal and phenobarbital induced rats. Biochem. Biophys. Res. Commun., 90, 531.

49) LECHNER, M.C., SINOGAS, C.M., manuscript in preparation.

50) NOKIN, P., HUEZ, G., MARBAIX, G., BURNAY, A., CHANTRENNE, H. (1976) Molecular Modifications associated with aging of globin messenger RNA in vivo. Eur. J. Biochem., 62, 509.

51) CARDELI, J., LONG, B., PITOT, H.C. (1976). Direct association of messenger RNA labelled in the presence of fluoro-orotate with membranes of the endoplasmic reticulum in rat liver. J. Cell Biol., 70, 47.

52) LANE, M.A., ADESNIK, M., SUMIDA, M., TASHIRO, Y., SABATINI, D.D. (1975). Direct association of messenger RNA with microsomal membranes in humam diploid fibroblasts. J. Cell. Biol., 65, 513.

53) BAR-NUN, S., KREIBICH, G., ADESNIK, M., ALTERMAN, L., NEGISHI, M., SABATINI, D.D. (1980). Synthesis and insertion of cytochrome P_{450} into endoplasmic reticulum membranes. Proc. Natl. Sci., 77, 965.

54) SHIRES, T.K., PITOT, H.C. (1974). The membron: a functional hypothesis for the translational regulation of genetic expression in Biomembranes, 5, 81. Ed. Lionel A. Manson. Plenum Press. New York - London.

55) LECHNER, M.C., SINOGAS, C.M. (1981). The importance of RNP's/ membrane interactions for stimulation of protein synthesis by phenobarbital. Biochem. Society Transact., 9, 156P.

56) McCAULEY, COURI, D. (1971). Early effects of phenobarbital on cytoplasmic RNA in rat liver. Biochim. Biophys. Acta, 238, 233.

57) COHEN, A.M., RUDDON, R.W. (1971). Stability of polyribosomes isolated from rat liver after phenobarbital administration. Mol. Pharmacol., 7, 484.

58) STEELE, W.J. (1970). Phenobarbital-induced prolongation of the half-life ribosomal-RNA of rat liver. Fed. Proc., 29, 737.

59) ORRENIUS, S., ERICSSON, J.L.E., ERNSTER, L. (1965). Phenobarbital-induced synthesis of the microsomal drug-metabolizing enzyme system and its relationship to the proliferation of endoplasmic membranes. A morphological and Biochemical study. J. Cell Biol., 25, 627.

60) SUNSHINE, G.H., WILLIAMS, D.J. RABIN, B.R. (1971). Role for steroid-hormones in interaction of ribosomes with endoplasmic membranes of rat-liver. Nature, Biol., 230, 133.

61) BRAZ, J., LECHNER, M.C. (1980). Nuclear poly(ADP-R) polymerase activity in rat liver during enzyme induction by phenobarbital. Effect of 5'-methyl-nicotinamide. Cienc. Biol. (Portugal), 5, 437.

62) BRAZ, J., LECHNER, M.C. (1980). Studies on poly(ADP-Ribose) polymerase activities in isolated nuclei from normal and phenobarbital induced rat livers. 1º Congr. Luso-Espanhol de Bioquimica P-157,23/26 Setembro,Coimbra-Portugal.

63) OGATA, N., UEDA, K., HAYAISHI, O. (1980). ADP-ribosylation of Histone H_2 B. Identification of glutamic acid residue 2 as the modification site. J. Biol. Chem., 255, 7610.

64) OGATA, N., UEDA, K., KAGAMIYAMA, H., HAYAISHI, O. (1980). ADP-ribosylation of Histone H_1. Identification of glutamic acid residues 2,14, and the COOH-terminal, lysine residue as modification sites. J. Biol. Chem., 255, 7616.

65) OKAYAMA, H., UEDA, K., HAYAISHI (1978). Purification of ADP-ribosylated nuclear proteins by covalent chromatography on dihydroxyboryl polyacrylamide beads and their characterization. Proc. Natl. Acad. Sci., USA, 75, 1111.

66) HAYAISHI, O., UEDA, K., OKAYAMA, H., KAWAICHI, M., OGATA, N., OKA, J., IKAI, K., ITO, S., SHIZUTA, Y., (1979). Poly (ADP-ribose) and ADP-ribosylation of proteins. in Enzymes, 151 Academic Press, Inc.

67) LEVIN, W., THOMAS, P.E., REIK, L., BRESNICK, E., RYAN, D.E. (1981). Characterization and regulation of rat liver microsomal cytochrome P_{450}. Biochem. Soc. Transaction, 9, 95P.

68) FUJII-KURIYAMA, Y., TANIGUCHI, T., MIZUKAMI, Y., SAKAY, M. TASHIRO, Y. , MURAMATSU, M. (1980). Molecular cloning of a

complementary DNA of phenobarbital-inducible Cytochrome P_{450} messenger RNA from the rat. Proc. Japan Acad., 56, Ser. B., 603.

69) MICHALOPOULOS, G., SATTLER, C.A., SATTLER, G.L., PITOT, H.C. (1976). Cytochrome P_{450} induction by phenobarbital and 3-methyl cholanthrene in primary cultures of hepatocytes. Science, 193, 907.

GENE MAPPING AND GENE TRANSFER IN MAMMALIAN CELLS

Frank H. Ruddle

Yale University
Department of Biology
New Haven, CT 06511

Somatic cell genetics provides a new approach to the genetic analysis of mammalian species. During the first decade of its use the rapid and simple methods of somatic cell genetics have been used to map over 200 genes in man (1,2). The total number now stands at more than 400. Somatic cell genetics depends on parasexual events detectable in somatic cells in vitro. These parasexual events result in the transfer of genetic information, and its fate in the recipient cell. More-over, these methods, in combination with current recombinant DNA technology, provide different levels of genetic resolution, ranging from the chromosomal location of genes to the fine structural mapping of individual genes.

Procedures of cell hybridization effect the transfer of entire nuclear genomes. Efficient methods for the formation of hybrid cells involves the use of membrane fusing agents such as inactivated Sendai virus (3) or polyethylene glycol (4). In order to efficiently recover hybrid clones, a variety of selective enrichment procedures are employed. The most common methods depend on the existence of different recessive and complementing auxotrophic markers in the contributing parental cell lines (5-7), but dominant markers have also been efficiently employed (8,9).

The ability to use hybrid cells for purposes of gene mapping derives from the spontaneous loss of one parent's chromosomes from the hybrid cell. Chromosome loss is initially rapid, but eventually the hybrid cells stably maintain a few donor chromosomes along with a full complement of the recipient

cell's chromosomes. In this way, a series of hybrid cell lines
may be derived, each containing a partial set of donor chromosomes. The factors determining which parental chromosome set is
lost are poorly understood, but human x mouse hybrids usually
segregate the human chromosome set.

In some instances, it is possible to experimentally
regulate the chromosome constitution of a hybrid cell. If a
donor chromosome carries a prototrophic gene that is under
selective pressure, only cells containing the gene and its
chromosome can survive. A good example involves the gene which
codes for the enzyme hypoxanthine-guanine phosphoribosyltransferase (HPRT), which is X-linked in all mammalian species
so far examined. Cells can be made dependent on HPRT activity
by placing them in medium containing hypoxanthine, aminopterin,
and thymidine (HAT medium). Thus, a mouse x human hybrid cell,
which was formed by fusing a mutant mouse cell deficient in
HPRT activity (HPRT$^-$) and a human cell prototrophic for HPRT
(HPRT$^+$), retains the human X chromosome when grown in HAT
medium. In addition, it is possible to select against cells
which have retained the X chromosome by exposing cells to toxic
metabolic analogues of hypoxanthine, such as thioguanine.

Given a means of distinguishing a donor gene product from
the homologous gene product of the recipient cell, it then
becomes possible to map genes. A series of hybrid cell lines
is examined for the expression of the phenotype of interest,
and expression is correlated with either the expression of
additional phenotypes (synteny testing), or the presence of a
particular chromosome (assignment testing). Once an assignment
is made, a gene can be localized to a region of the chromosome
by employing cell lines containing translocated or deleted
chromosomes (10) as one of the parents in hybrid formation. It
is then possible to correlate phenotype expression with the
portion of the translocated or deleted chromosome retained in
the hybrid (11). By examining hybrid cell lines obtained from
fusions with parental cell lines containing translocations or
deletions at varying points along the chromosome, it is possible
to localize genes to very precise regions of the chromosome.
Detailed descriptions of gene mapping strategies and procedures
have been published (1,7,12).

Cell hybridization methods have been used to map a large
number of genes in both man and mouse. The genes mapped have
generally been those coding for constitutive enzymes (13),
structural proteins (14), and cell surface antigens (15). These
types of genes are normally expressed in hybrid cells. More
recently, our laboratory has been involved in mapping genes

not normally expressed in hybrid cells, particularly genes which
code for developmentally specific functions. In these cases, the
sequence specificity of the DNA is used to distinguish donor and
recipient genes. Denatured DNA extracted from somatic cell
hybrids can be tested for its ability to form duplexes with a
donor-specific probe molecule by conventional Cot analysis (16).
Using a DNA probe to human globin, which cross-reacted very
weakly with mouse cell DNA, it was possible to screen a panel of
human x mouse hybrid cells for human α globin sequences and thus
map the gene to human chromosome 16 (17). Similarly, we were
able to assign the β globin gene to human chromosome 11 (18).
Neither the α or β globin genes were expressed in these hybrids,
and the map position of each gene was achieved by detecting the
presence of the human gene by nucleic acid hybridization and
correlating it with the presence of a human chromosome.

A second approach to the mapping of genes which are not
normally expressed in hybrid cells depends on the specificity
imparted by the distribution of restriction endonuclease sites.
High molecular weight DNA purified from a hybrid cell is
digested with a restriction endonuclease to produce fragments of
defined size. Whole cell DNA or DNA pre-fractionated by
reverse-phase chromatography (10), is then fractionated by
agarose gel electrophoresis and transferred to nitrocellulose
filters according to the blotting procedure of Southern (20).
Specific DNA fragments are identified in the blot by hybridization with a nucleic acid probe. This procedure is far more
efficient than liquid hybridization, as a single assay requires
approximately 50 µg of cellular DNA. Moreover, it is especially
useful in cases where the gene sequence of interest has been
conserved through evolution, as the homologous genes of different
species tend nevertheless to possess distinct restriction endonuclease sites. This results in different fragment patterns
following blotting, and allows the homologous genes of different
species to be distinguished.

We have demonstrated the effectiveness of the restriction
endonuclease mapping procedure using two different systems. In
both we have made use of cloned recombinant DNA molecules as
probes. First, we have confirmed the assignment of the β globin
gene to human chromosome 11 by associating the presence of globin
restriction fragments with the presence of human chromosome 11 in
human x mouse hybrids (21). Secondly, we have assigned the mouse
immunoglobulin kappa gene cluster to mouse chromosome 6 in an
analysis of mouse x Chinese hamster hybrids (22). It has become
clear to us from these studies that any cloned DNA sequence can be
mapped to a chromosomal site using this approach. These methods
will allow regulatory genes, introns, and structural gene
sequences to be mapped in the future. In addition, it will

be possible to map genes whose protein product is now readily
detectable or distinguishable between species, as long as an
appropriate nucleic acid probe is available.

The second somatic cell genetic method, microcell mediated
chromosome transfer, facilitates the transfer of only one or a
few donor chromosomes to a recipient cell (23,24). In this
procedure, donor cells are treated for a prolonged period of
time with a mitotic blocking agent. Eventually, the cells
escape the mitotic block to form polykaryocytes, in which the
chromosomes are divided among a number of micronuclei. When
these cells are treated with cytochalasin and subjected to
centrifugal force, the micronuclei are extruded yielding
microcells. The functional microcell possesses a micronucleus
containing one or a few chromosomes, a thin rim of cytoplasm,
and in enclosed by a plasma membrane. Although microcells
remain viable for only a few hours, their genetic information
can be efficiently rescued by fusion to a recipient cell
using either inactivated virus or polyethylene glycol as a
membrane fusing agent. Selection of microcell hybrids is
accomplished by choosing an appropriate auxotrophic recipient
cell, and by using purified microcell preparations. Employing
this strategy, a donor microcell chromosome bearing the complementing prototrophic gene will be present in all permutations.
The number of nonselected chromosomes introduced can be controlled
by choosing the size of the microcells employed for fusion.

The microcell transfer system possesses a number of
advantages over whole cell hybridization. First, the introduction
on only one or a few chromosomes into the recipient cell reduces
the complexity of the hybrid genome. This greatly simplifies
the correlation of a donor phenotype with a donor chromosome for
purposes of gene mapping. This situation is especially desirable
when mapping genes for which selective systems are available (25).
Second, hybridization of microcells introduces a minimal amount
of the donor cytoplasm into the recipient cell, thus reducing
the possibility of perturbing its epigenetic state. Finally,
the microcell procedure allows control over the direction of
chromosome segregation by choice of the microcell parent. This
aspect of the procedure eliminates one of the major limitations
of conventional cell hybridization and allows a number of new
systems to be studied. One of the most promising areas involves
human inherited diseases for which cultured cell lines are now
available (10). Somatic cell analysis of such diseases has been
hampered by the inability to use such cell lines as recipients
in a hybridization experiment. Microcells make it possible to
use mouse cells, or a variety of other species, to complement
the defect and define the genetic basis of such diseases. We

have made use of this approach in our genetic studies of
Xeroderma pigmentosum (26).

The third method of transfer, chromosome mediated gene
transfer (CMGT), was first described by McBride and Ozer (27).
This subject has been reviewed recently (28). Donor cells are
subjected to mitotic arrest and physically broken to release
metaphase chromosomes. The chromosomes are purified from nuclei
and cell debris, and applied to recipient cells in multiplicities of 0.5 to 2 genome equivalents per cell. When conditionally auxotrophic recipient cells (e.g., HPRT⁻) are treated
with chromosomes from a prototrophic donor, cells of the
recipient type that now express the donor phenotype
(transformants) can be recovered at low frequency. Two recent
technical improvements, co-precipitation of the chromosomes
with calcium phosphate and post-treatment of the recipient cells
with dimethyl sulfoxide, have allowed the frequency of transfer
to be raised as high as 2×0^{-5} transformants per recipient
cell (29,30). The size of the transferred genetic material
(referred to as the transgenome) is almost always subchromosomal in size. In most cases, it is impossible to detect
chromosomal material in transformants even though the transferred phenotype is expressed. In a minority of transformants
(10-15%) a cytologically detectable donor chromosome fragment
is present (29,30).

The transferred phenotype is initially expressed in an
unstable fashion. That is, between 3% and 10% of the cells
lose the transferred phenotype at each cell division. This
instability appears to be due to the actual loss of the transgenome, and not a modulation phenomenon, as it is impossible to
recover cells which reexpress the transferred phenotype following back selection (31). Stable cell lines spontaneously arise
from unstable transformants after an extended period of growth
in selective medium (32,33). In stable transformants, the
transgenome has become closely associated with recipient cell
chromosomes (24,34). In fact, in those cases where a detectable
donor chromosome fragment is present, it becomes a morpho-
logically distinct region of a recipient cell chromosome
following stabilization (35).

The CMGT procedure provides two means of deriving intergenic or subchromosomal breakage and disruption of linkage groups.
The first occurs during the uptake and integration of donor
chromosome material into the recipient cell. This is best
illustrated by CMGT experiments which have transferred the
human chromosome 17 gene coding for thymidine kinase (TK) (36).
The gene for the enzyme galactokinase (GALK) is closely
associated with TK on the long arm of human chromosome 17,

the two genes having been localized to a region representing
less than 0.2% of the human haploid genome. When TK is
selectively transferred in a CMGT, approximately 25% of the
transformants isolated are found to cotransfer for the gene for
GALK. Similarly, CMGT experiments transferring HPRT result in
transformants expressing the donor X-linked genes for phosphoglycerate kinase and the glucose-6-phosphate dehydrogenase in
some instances (22). The frequency of nonselected marker
cotransfer is inversely proportional to the genes' physical
distances from the selected gene. Cotransfer frequency thus
provides a quantitative measure of intergenic distance.

A second genetic diminution step in the CMGT process
occurs during the stabilization event. This was initially
observed as a loss of nonselected but cotransferred genes
following stabilization (34). We have recently demonstrated
how this step in the CMGT process may be used as a method for
determining the regional locations of genes on chromosomes (35).
An unstable TK transformant cell line, which had cotransferred
the linked human chromosome 17 genes for GALK and type I
procollagen (14), was used to generate a series of stable cell
lines. This cell line initially contained an independent
cytologically detectable chromosome fragment identical to the
long arm of human chromosome 17. In each of the stable subclones the donor chromosome fragment had become associated with
a recipient cell chromosome. The size of the fragment, however,
decreased following stabilization, with varying portions of the
long arm of human chromosome 17 being retained in the stable
subclones. Furthermore, expression of the nonselected donor
genes was correlated with the size of the fragment retained.
The effective result was a deletion map which allowed ordering
and localization of the human chromosome 17 genes. Use of
chromosome fragments and the stabilization process thus provides
a qualitative mapping tool, complementing the quantitative
mapping data obtained by cotransfer frequency. These two
methods combine to allow precise intergenic mapping.

In recent years, it has become possible to genetically
transform somatic cells in vitro (36) and mammalian embryos (37)
using purified DNA. This field has been recently reviewed
(38,39,40).

An additional method of genetic analysis involves the
direct injection of RNA sequences into cultured cells. The cell
then serves as an in vivo translation system. Coupling of this
procedure with a method of detecting a particular gene product
provides an assay for the nucleic acid sequence governing its
synthesis. Our laboratory has recently developed an effective
mRNA injection technique using cultured mammalian cells as a

recipient (42). Whole cell RNA has been isolated from the primary human fibroblast cell line FS11 (43), and polyadenylated mRNA was obtained by passage over an oligo-(dT) cellulose column (44). The mouse L-cell LTK$^-$, deficient in both TK and adenine phosphoriboysltransferase (APRT) activity, has been used as a recipient. When injected with the FS11 mRNA by micro-manipulative technique (45), the LTK$^-$ cell line transiently expresses both TK and APRT activity measured by the cell's ability to incorporate radioactively labelled thymidine and adenine. This technique has also proved effective for the transfer of specialized gene functions such as interferon production (42).

ACKNOWLEDGMENTS

Thanks is given to my colleagues for their criticisms and suggestions. Special thanks goes to Mrs. Marie Siniscalchi for the preparation of the manuscript. Much of this work has been supported by NIH grant GM09966.

REFERENCES

1. McKusick, V. A. and F. H. Ruddle. 1977. Science 196: 390-405.
2. Donald, L. J. and J. L. Hamerton. 1978. Cytogenet. Cell Genet. 22: 5-11.
3. Harris, H. and J. F. Watkins. 1965. Nature 205: 640-646.
4. Pontecorvo, G. 1975. Somatic Cell Genet. 1: 397-400.
5. Littlefield, J. 1964. Science 145: 709-710.
6. DeMars, R. 1974. Mutat. Res. 24: 335-364.
7. Creagan, R. P. and F. H. Ruddle. 1977. In: Molecular Structure of Human Chromosomes (J. Yunis, ed.), pps. 89-142, Academic Press, New York.
8. Baker, R. M., D. M. Brunette, R. Mankovitz, L. H. Thompson, G. F. Whitmore, L. Siminovitch and J. E. Till. 1974. Cell 1: 9-21.
9. Siminovitch, L. 1976. Cell 7: 1-11.
10. Greene, A. E. 1978. List of genetic variants, chromosomal aberrations and normal cell cultures submitted to the repository -- Oct. 1980. Available from the Hum. Genet. Mutant Cell Repos., Camden, New Jersey.
11. Ricciuti, F. and F. H. Ruddle. 1973. Nature New Biol. 241: 180-182.
12. Ruddle, F. H. and R. P. Creagan. 1975. In: Annual Review of Genetics (H. L. Roman, A. Campbell and L. M. Sandler, eds.), vol. 9, pps. 407-486, 1975. Ann. Review, Inc., Palo Alto, CA.
13. Nichols, E. A. and F. H. Ruddle. 1973. J. Histochem. Cytochem 21: 1066-1081.

14. Sundar Raj, C. W., R. L. Church, L. A. Klobutcher and
         F. H. Ruddle. 1977. Proc. Natl. Acad. Sci. USA 74:
         4444-4448.
15. Dorman, B. P., N. Shimizu and F. H. Ruddle. 1976.
         J. Cell Biol. 70: 77a.
16. Britten, R. J. and D. E. Kohne. 1968. Science 161:
         529-540.
17. Deisseroth, A., A. Nienhuis, P. Turner, R. Velez, W. French
         Anderson, F. H. Ruddle, J. Lawrence and
         R. Kucherlapati. Cell 12: 205-218.
18. Deisseroth, A., A. Nienhuis, J. Lawrence, R. Giles,
         P. Turner and F. Ruddle. 1978.
         Proc. Natl. Acad. Sci. USA 75: 1465-1460.
19. Seidman, J. G., A. Leder, M. Nau, B. Norman and P. Leder.
         1978. Science 202: 11-17.
20. Southern, E. M. 1975. J. Mol. Biol. 98: 503-517.
21. Huttner, K. M., G. A. Scangos and F. H. Ruddle,
         unpublished results.
22. Swan, D., P. D'Eustachio, L. Leinwand, J. Seidman,
         D. Keithley and F. Ruddle. 1979.
         Proc. Natl. Acad. Sci. USA 76: 2735-2739.
23. Fournier, R.E.K. and F. H. Ruddle. 1977a.
         Proc. Natl. Acad. Sci. USA 74: 319-323.
24. Fournier, R.E.K. and F. H. Ruddle. 1977b.
         Proc. Natl. Acad. Sci. USA 74: 3937-3941.
25. Kozak, C. A., R.E.K. Fournier, L. A. Leinwand and
         F. H. Ruddle. 1979. Biochem. Genet. 17: 23-34.
26. Lin, Pin-fang and F. H. Ruddle. 1981. Nature 289:
         191-194.
27. McBride, O. W. and H. L. Ozer. 1973.
         Proc. Natl. Acad. Sci. USA 70: 1258-1262.
28. Klobutcher, L. A. and F. H. Ruddle. Ann. Rev. Biochem.
         50: 533-554, 1981.
29. Miller, C. L. and F. H. Ruddle. 1978.
         Proc. Natl. Acad. Sci. USA 75: 3346-3350.
30. Klobutcher, L. A., C. L. Miller and F. H. Ruddle. 1980.
         Proc. Natl. Acad. Sci. USA 77: 3610-3614.
31. Willecke, K. and F. H. Ruddle. 1975.
         Proc. Natl. Acad. Sci. USA 72: 1792-1796.
32. Degnen, G. E., I. L. Miller, E. A. Adelberg and
         J. M. Eisenstadt. 1977. Proc. Natl. Acad. Sci. USA
         74: 3956-3959.
33. Athwal, R. and O. W. McBride. 1977.
         Proc. Natl. Acad. Sci. USA 74: 2943-2947.
34. Willecke, K., R. Lange, A. Kruger and T. Reber. 1976.
         Proc. Natl. Acad. Sci. USA 73: 1274-1278.
35. Klobutcher, L. A. and F. H. Ruddle. 1979.
         Nature 280: 657-660.
36. Scangos, G. A., K. M. Huttner, D. K. Juricek and
         F. H. Ruddle. 1981. Molec. Cell. Biol. 2: 111-120.

37. Gordon, J. W., G. A. Scangos, D. J. Plotkin, J. A. Barbosa
    and F. H. Ruddle. 1980.
    Proc. Natl. Acad. Sci. USA 77: 7380-7384.
38. Scangos, G. A. and F. H. Ruddle. 1981. Gene, in press.
39. Huttner, K. M. and F. H. Ruddle. 1981. J. Cancer Inst.,
    in press.
40. Ruddle, Frank H. 1981. Nature, in press.
41. Ruddle, F. H. and O. W. McBride. 1977. In: The Molecular
    Biology of the Mammalian Genetic Apparatus
    (P.O.P. Ts'o, ed.), chap. 14, pps. 163-169.
    ASP Biological and Medical Press, Amsterdam, 1977.
42. Liu, C. P., D. L. Slate and F. H. Ruddle. 1979.
    Proc. Natl. Acad. Sci. USA 76: 4503-4506.
43. Deeley, R. G., J. I. Gordon, A. T. Burns, K. P. Mullinix,
    M. Binastein and R. F. Goldberger. 1977.
    J. Biol. Chem. 252: 8310-8319.
44. Alberts, B. and G. Herrick. 1971. In: Methods in
    Enzymology (K. Moldave and L. Grossman, eds.),
    Vol. XXI, pps. 198-217, Academic Press, New York.
45. Diacumakos, E. G. 1973. In: Methods in Cell Biology
    (D. M. Prescott, ed.), pps. 287-311,
    Academic Press, New York.

DIHYDROFOLATE REDUCTASE GENE AMPLIFICATION, ALTERED DIHYDROFOLATE
REDUCTASE, AND METHOTREXATE RESISTANCE IN CULTURED 3T6 CELLS
ASSOCIATED WITH UNSTABLE AMPLIFICATION OF AN ALTERED DIHYDROFOLATE
REDUCTASE GENE

Robert T. Schimke and Daniel A. Haber

Department of Biological Sciences
Stanford University
Stanford, California 94305

Gene Amplification and Methotrexate Resistance in Cultured Animal
Cells

Resistance of human neoplasms to various cancer chemothera-
peutic agents has been a frustrating problem. Extensive studies
in many laboratories have investigated the development of
resistance to the 4-amino analog of folate, methotrexate (MTX) and
have defined three general mechanisms for such resistance, including
reduction in MTX transport,[1] alteration in affinity of MTX for
the target enzyme of MTX inhibition, dihydrofolate reductase
(DHFR)[2-5] and increased levels of DHFR.[7-13] Our laboratory has
shown that the elevated DHFR enzyme levels in various MTX-resistant
cell lines result from a corresponding increase in the number of
DHFR genes, i.e., gene amplification[6,14] irrespective of whether
the karyotype is grossly aneuploid or is relatively stable, and
irrespective of whether the MTX-resistance is phenotypically
stable or unstable.

Gene amplification as a mechanism for the acquisition of drug
resistance in cultured animal cells is not limited to MTX. Wahl
et al. [15] have shown that resistance to an inhibition of aspartyl-
transcarbamylase (PALA) results from amplification of a DNA
sequence coding for this enzyme. In addition, Beach and Palmiter[16]
have ascribed resistance to cadmium in mouse L1210 cells to amplification of the metallothionine gene. In addition there are
various examples of stepwise selection for high drug resistance
resulting in increases in specific enzymes,[17-21] and it is likely
that certain of these examples are the result of selective gene
amplification. We suggest that amplification in cultured animal

97

cells may be relatively frequent, and they may not have been
detected because of the necessity for stepwise selection, as well
as the fact that, more often than not, the initial occurrence is
an unstable event.

Stable and Unstable Amplification and Localization of Amplified Dihydrofolate Reductase Genes

When cells are grown in the absence of MTX, the MTX-resistance
phenotype and DHFR gene amplification can be either stable or
unstable. In the latter case approximately 50% of the amplified
genes are lost within 20 cell doublings.[13,22] In all cases of
stable amplification of DHFR genes we have studied,[23-25] the
amplified genes are present on chromosomes. In general, the
amplified genes are associated with an expanded region of a single
chromosome which is denoted a homogeneously staining region (HSR)[26]
on the basis of the lack of normal trypsin-giemsa banding patterns.
Characteristically only one of the two homologous chromosomes has
a HSR. In Chinese hamster ovary cell lines,[23] as well as Chinese
hamster lung cell lines[26] the HSR is present on the long arm of
chromosome 2, and the non-amplified DHFR gene has recently been
mapped to this region.[27] The implication is that in such cases
of chromosomal amplification, the amplified genes occur at the
site of the resident (non-amplified) gene.

In all unstably amplified cell lines we have studied,[13,22,25]
the DHFR genes are present on extrachromosomal elements, called
double minute chromosomes (DMs). These elements are self-
replicating and do not contain centromeric staining regions.[28]
As a consequence of their not participating in the process of
equal segregation of chromosomes at mitosis, they can be distributed unequally into daughter cells. We have shown that cells
with lower numbers of DMs grow more rapidly than cells with large
numbers of DMs.[29] Thus, when cells are grown in the absence of
MTX, those cells in which progressively fewer DMs are distributed
will become dominate in the cell population. Another mechanism
for loss of extrachromosomal, amplified DHFR genes involves the
process of micronucleation whereby nuclear membrane reassembles
around aggregates of DMs, producing packets of what appear as
cytoplasmic "micronuclei", and which are subsequently lost from
cells, resulting in extremely rapid loss of large numbers of
amplified genes.

It is interesting to note that mouse cell lines characteristically generate MTX-resistant variants that are unstably
amplified, whereas hamster cell lines generate MTX-variants with
stably amplified genes.[30] Whether this relates to the difference
in stable karyotypes (characteristic of hamster cell lines) or of

differences in DNA sequences within or surrounding the DHFR gene
in the different species is unknown.

Unstable Amplification on an Altered Dihydrofolate Reductase Gene

As mentioned previously, MTX resistance can result from
different mechanisms. Flintoff and Siminovitch[3] mutagenized CHO
cells and obtained variants with resistance resulting from a
5-fold decrease in the affinity of MTX for DHFR. Upon further
selection at higher MTX concentrations, they obtained cells with
increased amounts of the altered enzyme, presumably as a result
of amplification of the altered gene. In our laboratory we have
been studying the properties of a mouse fibroblast 3T6 cell line
resistant to 400 µM MTX, i.e., 3T6-R400. This cell line was
selected for stepwise MTX resistance by Drs. V. Morhenn and R.
Kellems with intermediate cell lines denoted 3T6-R5 and 3T6-R50.
We have studied the 3T6-R50 cell line extensively.[22] Its
resistance to 50 µM MTX is unstable, and DMs are lost upon growth
in the absence of MTX. This cell line contains approximately 50
copies of the DHFR gene on DMs. The 3T6-R400 cell line was
derived from the 3T6-R50, and was first examined in our laboratory
after growth in 400 µM MTX continuously for approximately 18 months.
When examined employing the fluorescence activated cell sorter and
the fluorescein-conjugate of MTX which quantitates the amount of
DHFR per cell,[31] the 3T6-R400 showed fluorescence indistinguishable
from that of sensitive 3T6 cells, and this pattern was clearly
distinguishable from that of the 3T6-R50 cells which showed
fluorescence indicative of increased amounts of DHFR (Fig. 1,a-c).
The minimal fluorescence of the 3T6-R400 cells results from the
fact that these cells contain large amounts of a DHFR that is
drastically altered in its kinetic properties.[32] These properties
include: (1) a K_i for MTX that is 300 times less than the K_i for
the wild-type DHFR; (2) a V_{max} that is 1/20th that of normal DHFR;
(3) heat stability similar to normal DHFR; (4) a K_m for NADPH
that is similar to normal DHFR; (5) a K_m for dihydrofolate that
is three times that of normal DHFR; (6) a size similar to normal
DHFR but a charge difference (more basic). The 3T6-R400 cells
contain approximately 30 copies of the DHFR gene, these genes
reside on DMs, and such resistance is unstable. Thus, the reason
the cells do not fluoresce is because of the low affinity of
fluorescein-MTX for DHFR.

The question we have posed is: Did the mutation occur on
the original chromosomal DHFR and subsequently undergo amplification, or did the mutation occur subsequently to amplification,
and DMs with altered genes become predominant in cells, and in
the cell population? The latter explanation suggests that cells
with an altered DHFR gene present on DMs have a selective advantage

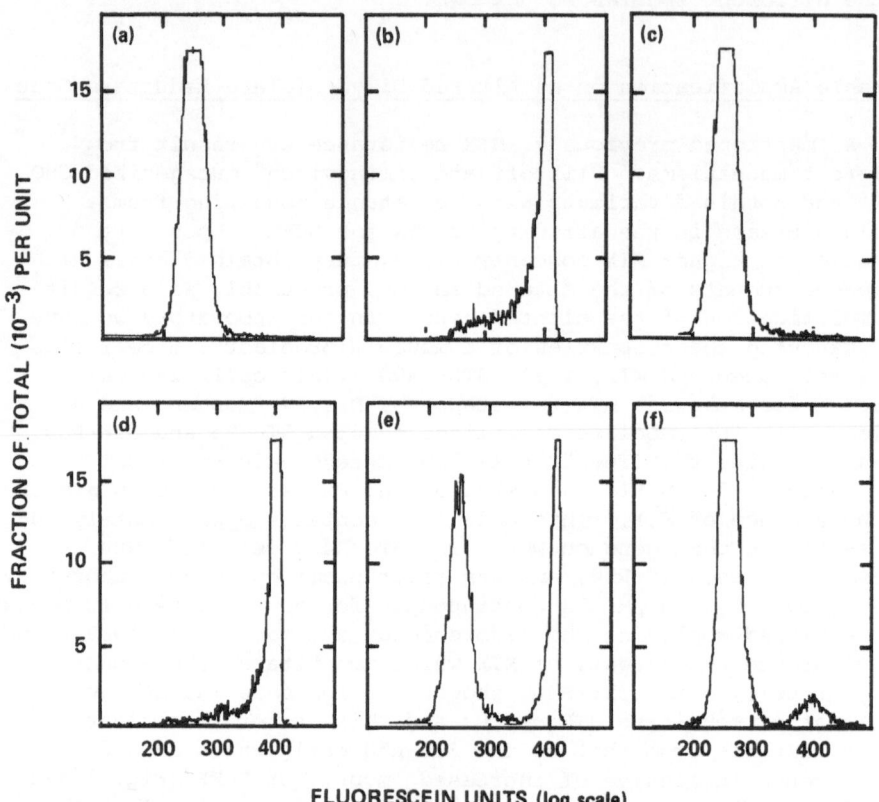

FLUORESCEIN UNITS (log scale)

Fig. 1. Fluorescence of cells grown in progressively increasing
MTX concentrations. Cultures were grown in the absence of MTX for
at least 3 generations and were then incubated for 20 hours in 30
μM MTX-F, glycine, hypoxanthine and thymidine. The cells were
trypsinized, centrifuged through a 0.3M sucrose cushion and examined
using the fluorescence activated cell sorter (FACS). Fluorescence
is in arbitrary units with 400 fluorescence units representing
maximal fluorescence. Panel A represents the fluorescence of
parental, MTX-sensitive cells; Panel B, that of 3T6-R50 cells;
Panel C, that of 3T6-R400 cells; Panel D, that of 3T6-R50 cells
grown for 60 generations at 400 μM MTX. Panel E shows a 1:1
mixture of 3T6-R400 cells and 3T3-R500 cells. Panel F represents
the fluorescence profile of this mixed population after 50 generations of growth at 400 μM MTX. Fluorescence is plotted on a log
scale in order to compare the minimal fluorescence of MTX-sensitive
cells and cells expressing primarily altered DHFR with the high
fluorescence of cells containing elevated levels of normal enzyme.
                    (Haber and Schimke, to be published)

when grown in 400 μM MTX. That such is the case is shown in Fig.
le,f where we have mixed equal numbers of cells resistant to 400
μM MTX, the 3T6-R400 cells, and 3T3-R500 cells which contain large
amounts of normal DHFR.   In as short a time as 60 generations, the
3T6-R400 cells constitute 95% of the cells growing at 400 μM MTX,
indicating that cells with altered enzyme have a growth advantage.

If an occasional cell with altered DHFR genes were present in
the progenitor 3T6-R50 cell population, then growth of these cells
at 400 μM MTX should reveal its presence. Fig. ld shows that when
these cells were grown at 400 μM MTX for 60 generations, the cell
population emerging contained no cells with fluorescence characteristics of the 3T6-R400 cells. Thus we conclude that the progenitor
population contained no such cells.

Our next series of experiments was designed to determine if
the altered DHFR gene resided on a chromosome, or if all of the
altered DHFR genes resided on DMs. Thus, we grew cells in the
absence of MTX for long periods of time. In order to detect the
altered enzyme, we determined the effect of MTX on inhibition of
the enzyme. Fig. 2 shows such inhibition profiles. Cells grown
in 400 μM MTX showed an ID_{50} of approximately 6×10^{-7}. Because
of the large amount of altered enzyme in spite of its low V_{max},
no normal enzyme (ID_{50} of 1×10^{-9}) could be detected. After 25
generations in MTX-free medium, some normal DHFR could be detected,
and after 60 generations in MTX-free medium, the inhibition profile
was similar to that of the normal enzyme. This loss of altered
DHFR was associated with the loss of DMs from metaphase spreads
of these cells. Inasmuch as the normal enzyme was detectable in
these cells, we conclude that the normal gene has not been lost.
This analysis was undertaken with the 3T6-R400 cell population,
as well as with 3 clones from that population and the results were
similar. After a total of 110 generations in MTX-free medium, we
subjected the cells to stepwise selection in MTX, and when they
were resistant to 100 μM MTX, we determined their fluorescence
pattern. We found that 90% of the cells showed fluorescence
patterns (Fig. 1) characteristic of the altered DHFR, whereas 10%
showed amplification of the normal DHFR. Although such results
suggest that the altered DHFR resides on a chromosome, it is
well-known that DMs tend to stick to chromosomes, and are
difficult to remove completely from cells as a result of unequal
distribution into daughter cells. Consequently, we derived after
the 110 generations, 6 clones from the original clone #2 of the
3T6-R400, a process that required an additional 20 generations.
When these cells were subjected to MTX selection, 2 clones amplified the altered DHFR gene, whereas 4 amplified the normal enzyme.
At the time that the 6 subclones were obtained each of these
subclones was itself sub-cloned, a process requiring another
additional 20 generations. Several of these were examined as

well. Thus, of 5 sub-subclones of a subclone that generated
altered enzyme resistance, only 1 generated resistance with altered
DHFR, whereas 4 had amplified the normal DHFR gene. Of 10 sub-
subclones of a subclone that generated MTX resistance with normal
DHFR amplification, all 10 amplified the normal DHFR gene. The
biology of the MTX selection was also different between cells
amplifying altered and normal DHFR. Cells amplifying the altered
DHFR gene grew readily at each selection step, whereas cells
amplifying the normal DHFR required long time periods for the
generation of MTX-resistance cells.

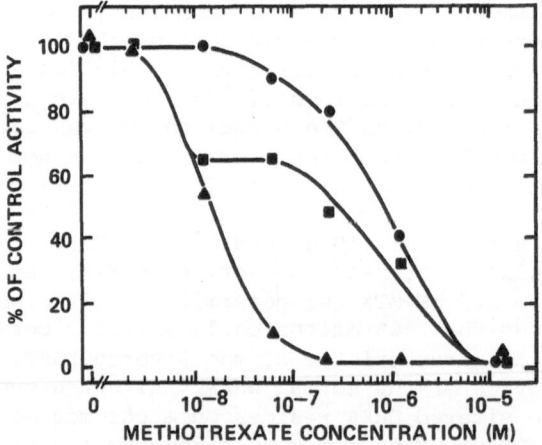

Fig. 2. Inhibition by MTX of DHFR activity from 3T6-R400 cells as
a function of growth out of selection. 3T6-R400 cells were grown
in the absence of MTX for zero (O), 25 (□), and 60 (Δ)
generations. Soluble cellular extract was prepared from these and
DHFR activity was measured as a function of varying MTX concentrations. Activity is expressed as a percent of uninhibited levels,
which were comparable for the three extracts. The ID_{50} values
obtained by this analysis (6×10^{-9}M MTX for the normal DHFR and
6×10^{-7} for the altered enzyme) are useful for comparison purposes,
but are not a true measure of MTX binding affinity (see Ref. 32).

We conclude from these experiments that if a cell contains an altered DHFR gene, it can be amplified readily, and resistance develops readily. In contrast, when cells have lost the last DM, only the normal DHFR gene can be amplified. These results are consistent with the conclusion that all altered DHFR genes reside on DMs, and that the chromosomal DHFR genes are all of the normal type.

Inasmuch as the 3T6-R400 cells we have studied were grown in 400 µM MTX for 18 months, we sought a sample of these cells shortly after initial selection at 400 µM MTX. Fortunately, Dr. V. Morhenn, who originally generated these cell lines, had derived several clones of these cells. We have examined one such clone (denoted clone 5). A fluorescence analysis of Fig. 3a showed that the majority of these cells had fluorescence properties similar to the cells with altered DHFR. However, a few cells had high fluorescence. We sorted cells with greater than 350 fluorescein units/cell and grew them in 400 µM MTX for 25 generations, generating a clearly apparent population with high fluorescence. From such cells (Fig. 3b) we again sorted cells with greater than 350 fluorescein units/cell. Fig. 3 c-e shows the fluorescence pattern of these cells grown 15, 40, and 60 generations in 400 µM MTX. After 15 generations the pattern is vastly heterogeneous, and progressively the emerging cells are those with a predominance of altered enzyme (and genes). Fig. 3f shows a predominately minimal fluorescence sorted subpopulation of the original 3T6-R400 clone 5, and when these cells were grown in 400 µM MTX for 50 generations, there emerged a few cells with high fluorescence characteristics (Fig. 3g).

We propose that by isolating the highly fluorescent subpopulation of the clone 5 cells, we obtained a cell population similar to that which prevailed shortly after the appearance of the altered DHFR gene in the MTX-resistant 3T6-R400 cells. These cells contain numerous unstably amplified normal DHFR genes, as well as a few DMs with altered DHFR genes, and subsequent growth of these cells continuously in MTX results in the selective outgrowth of cells with an increasing proportion of altered genes. Such an explanation is consistent with the fact that cells with fewer DMs will grow more rapidly and, at a given MTX concentration, i.e., 400 µM, cells with altered DHFR genes will become predominant inasmuch as fewer altered DHFR enzyme molecules (and DHFR genes) are required, as a result of the markedly decreased affinity for MTX.

From these studies we make the following observations and conclusions.

1. Mutations affecting MTX binding to DHFR are likely to be very common. This follows from the fact that there are multiple sites

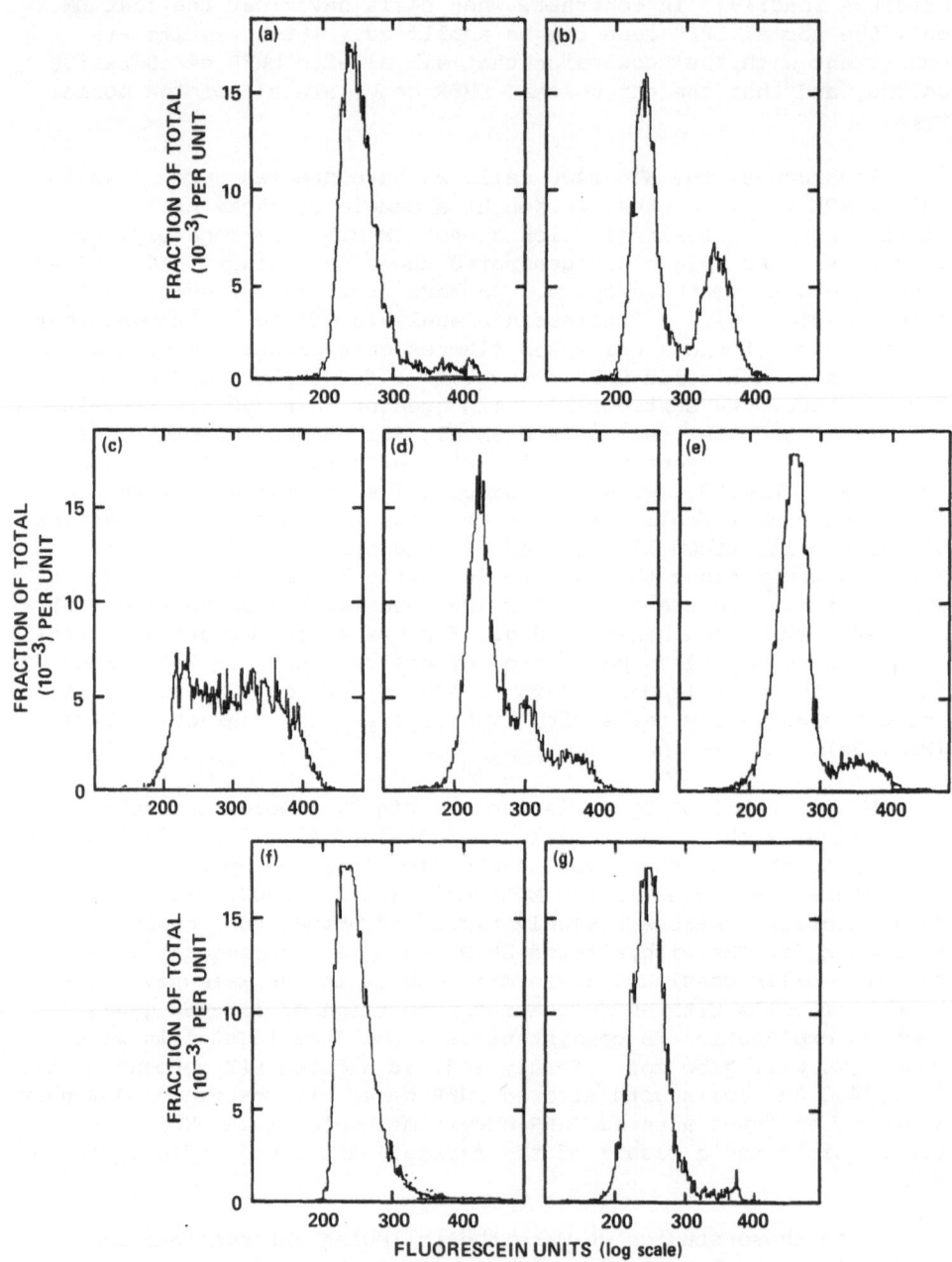

FLUORESCEIN UNITS (log scale)

involved in the binding of MTX to DHFR[33] and mutations in a number of these amino acids can decrease MTX affinity. Thus, it is not surprising that various laboratories have found altered DHFR with different kinetic properties.[2-5,32]

2. Our ability to readily detect the altered enzyme results from the fact that the altered DHFR gene occurs as an extrachromosomal element, and hence the predominance of this gene within cells is subject to the unequal distribution of genes and the growth advantage of cells with fewer extrachromosomal elements. If in contrast, a mutation had occurred among one of a number of chromosomal localized DHFR genes, which are not subject to such phenomena of unequal distribution, it is unlikely that it would be detected readily. Thus, it is possible that in cell lines with stably integrated DHFR genes, there may occur one or more altered DHFR genes, and hence the DHFR present in the cells may constitute more than one species.

3. It is unlikely that a vastly altered DHFR could first arise as a result of a chromosomal mutation in the DHFR gene. Thus, if the mutation occurred on a chromosome, and the cells were selected for low MTX resistance, the normal DHFR allele (enzyme) would be completely inhibited by MTX, and the residual, altered DHFR enzyme with 1/20 the catalytic activity would not be capable of generating sufficient tetrahydrofolate for cell survival. Rather such vastly altered genes most likely occur only when there is sufficient amplification of the DHFR gene so that when the altered gene is present in a highly amplified state, the total

---

Fig. 3. Isolation of highly fluorescent and minimally fluorescent subpopulations of 3T6-R400 clone 5 cells. The fluorescence pattern of an early subclone of 3T6-R400 cells (clone 5) was examined (Panel A). The small population of brightly fluorescent cells (above 350 fluorescence units) was isolated and grown in 400 µM MTX for 25 generations (Panel B). The brightly fluorescent peak was again isolated and grown in 400 µM MTX, this time for only 15 generations (Panel C). These cells were then grown for a total of 40 generations (Panel D) and 60 generations (Panel E) in 400 µM MTX. The predominant minimally fluorescent subpopulation of the original 3T6-R400 clone 5 cells (Panel A) was also isolated (Panel F). After 50 generations in 400 µM MTX, these cells were again analyzed by FACS (Panel G). (Haber and Schimke, to be published.)

amount of DHFR catalytic capacity, i.e., a large number of enzyme molecules with a low turnover number, is sufficient to generate the necessary tetrahydrofolate.

4. The double minutes contain only functionally active altered DHFR genes. If the DMs contained a normal and an altered DHFR gene, then cells with multiple DMs should contain equal proportions of normal and altered DHFR, and this is not the case (Fig. 2).

5. Double minute chromosomes are self-replicating. As long as there is a single DM per cell, the progeny of such cells rapidly distribute the replicated DMs unequally into daughter cells, thereby allowing for the rapid emergence of MTX resistance. When all DMs are lost, then the subsequent amplification of a chromosomal gene is a rare event.

The presence in cultured mammalian cells of self-replicating extrachromosomal DNA sequences which provide a selective growth advantage and whose presence and persistence in cells may be under control mechanisms other than those of classical genetics adds a new dimension to somatic cell genetics. Possible mechanisms for amplification are discussed elsewhere.[34]

REFERENCES

1. F. M. Sirotnak, S. Kurita, and D. J. Hutchison, On the nature of a transport alteration determining resistance to amethopterin in the L1210 leukemia. Cancer Research 28:75 (1968).

2. A. M. Albrecht, J. L. Biedler, and D. J. Hutchison, Two different species of dihydrofolate reductase in mammalian cells differentially resistant to amethopterin and methasquin. Cancer Research 32:1539 (1972).

3. W. F. Flintoff, S. V. Davidson, and L. Siminovitch, Isolation and partial characterization of three methotrexate-resistant phenotypes from Chinese hamster ovary cells. Somatic Cell Genetics 2:245 (1976).

4. R. C. Jackson and D. Niethammer, Acquired methotrexate resistance in lymphoblasts resulting from altered kinetic properties of dihydrofolate reductase. Eur. J. Cancer 13:567 (1977).

5. J. H. Goldie, G. Krystal, D. Hartley, G. Gudauskas, and S. Dedhar, A methotrexate insensitive variant of folate reductase present in two lines of methotrexate-resistant L5178Y cells. Europ. J. Cancer 16:1539 (1980).

6. M. T. Hakala, S. F. Zakrzewski, and C. A. Nichol, Relation of folic acid reductase to amethopterin resistance in cultured mammalian cells. J. Biol. Chem. 236:952 (1961).

7. F. W. Alt, R. E. Kellems, J. R. Bertino, and R. T. Schimke, Selective multiplication of dihydrofolate reductase genes in methotrexate-resistant variants of cultured murine cells. J. Biol. Chem. 253:1357 (1978).

8. R. C. Jackson, L. I. Hart, and K. R. Harrap, Intrinsic resistance to methotrexate of cultured mammalian cells in relation to the inhibition kinetics of their dihydrofolate reductases. Cancer Research 36:1991 (1976).

9. W. F. Flintoff, and K. Essani, Methotrexate-resistant Chinese hamster ovary cells contain a dihydrofolate reductase with an altered affinity for methotrexate. Biochem. 19:4321 (1980).

10. R. P. Raunio, and M. T. Hakala, Comparison of folate reductases of sarcoma 180 cells, sensitive and resistant amethopterin. Mol. Pharmacol. 3:279 (1967).

11. H. Nakamura and J. W. Littlefield, Purification, properties, and synthesis of dihydrofolate reductase from wild type and methotrexate-resistant hamster cells. J. Biol. Chem. 247:179 (1972).

12. U. J. Hanggi and J. W. Littlefield, Altered regulation of the rate of synthesis of dihydrofolate reductase in methotrexate-resistant hamster cells. J. Biol. Chem. 251:3075 (1976).

13. F. W. Alt, R. E. Kellems, and R. T. Schimke, Synthesis and degradation of folate reductase in sensitive and methotrexate-resistant lines of S-180 cells. J. Biol. Chem. 251:3063 (1976).

14. R. T. Schimke, R. J. Kaufman, J. H. Nunberg, and S. L. Dana, Studies on the amplification of dihydrofolate reductase genes in methotrexate resistant cultured mouse cells. Cold Sprg. Harbor Symp. Quant. Biol. 43:1297 (1979).

15. G. M. Wahl, R. A. Padgett, and G. R. Stark, Gene amplification causes over production of the first three enzymes of UMP synthesis in N-(phosphoacetyl 1-aspartate)-resistant hamster cells. J. Biol. Chem. 254:8679 (1979).

16. L. R. Beach and R. D. Palmiter, Amplification of the metallothionein-1 gene in cadmium-resistant mouse cells. Proc. Natl. Acad. Science, USA (1981) in press.

17. F. Baskin, S. C. Carlin, P. Kraus, M. Friedkin, and R. N. Rosenberg, Experimental chemotherapy of neuroblastoma. II. Increased thymidylate synthetase activity in a 5-fluorodeoxyuridine-resistant variant of mouse neuroblastoma. Molecular Pharma. 11:105 (1975).

18. M. Sinensky, Isolation of a mammalian cell mutant resistant to 25 hydroxycholesterol. Biochem. Biophys. Res. Comm. 78:863 (1977).

19. M. Meuth and H. Green, Alterations leading to increased
     ribonucleotide reductase in cells selected for resistance
     to deoxynucleosides. Cell 3:367 (1974).
20. F. Baskin, R. Rosenberg, and V. Dev, Correlation of double
     minute chromosomes with unstable multi-drug cross-
     resistance in nueroblastoma uptake mutants. Proc. Natl.
     Acad. Sci., USA (1981) in press.
21. J. A. Wright, W. H. Lewis, and C. L. J. Parfett, Somatic cell
     genetics: A review of drug resistance, lectin resistance
     and gene transfer in mammalian cells in culture. Can. J.
     Genet. Cytol. 22:443 (1980).
22. P. C. Brown, S. M. Beverley, and R. T. Schimke, Relationship
     of amplified dihydrofolate reductase genes to double minute
     chromosomes in unstably resistant mouse fibroblast cell
     lines. Molecular and Cellular Biol. (1981) in press.
23. J. N. Nunberg, R. J. Kaufman, R. T. Schimke, G. Urlaub, and
     L. A. Chasin, Amplified dihydrofolate reductase genes are
     localized to a homogeneously staining region of a single
     chromosome in a methotrexate resistant Chinese hamster
     ovary cell line. Proc. Natl. Acad. Sci., USA 75:5553 (1978).
24. B. J. Dolnick, R. J. Berenson, J. R. Bertino, R. J. Kaufman,
     J. H. Nunberg, and R. T. Schimke, Correlation of dihydro-
     folate reductase elevation with gene amplification in a
     homogeneously staining chromosomal region of L5178Y cells.
     J. Cell Biol. 83:394 (1979).
25. R. J. Kaufman, P. C. Brown, and R. T. Schimke, Amplified
     dihydrofolate reductase genes in unstably methotrexate-
     resistant cells are associated with double minute chromo-
     somes. Proc. Natl. Acad. Sci., USA 76:5669 (1979).
26. J. L. Biedler and B. A. Spengler, Metaphase chromosome anomaly:
     Association with drug resistance and cell-specific products.
     Science 191:185 (1976).
27. M. Roberts, K. M. Huttner, R. T. Schimke, and F. H. Ruddle,
     Chromosomal assignment for the native Chinese hamster
     dihydrofolate reductase gene. J. Cell Biol.(1980) in press.
28. P. E. Barker and T. C. Hsu, Double minutes in human carcinoma
     cell lines, with special reference to breast tumors.
     J. Natl. Cancer Inst. 62:257 (1979).
29. R. J. Kaufman, P. C. Brown, and R. T. Schimke, Loss and
     stabilization of amplified dihydrofolate reductase genes
     in mouse sarcoma S-180 cell lines. Molecular and Cellular
     Biol. (1981) in press.
30. R. J. Kaufman and R. T. Schimke, Amplification and loss of
     dihydrofolate reductase genes in a Chinese hamster ovary
     cell line. Molecular and Cellular Biol. (1981) in press.
31. R. J. Kaufman, J. R. Bertino, and R. T. Schimke, Quantitation
     of dihydrofolate reductase in individual parental and
     methotrexate-resistant murine cells. J. Biol. Chem.
     253:5852 (1978).

32. D. A. Haber, S. M. Beverley, M. L. Kiely, and R. T. Schimke, Properties of an altered dihydrofolate reductase encoded by amplified genes in cultured mouse fibroblasts. J. Biol. Chem. (1981) in press.

33. D. A. Matthews, R. A. Alden, S. T. Bolin, S. T. Freer, R. Hamlin, N. Xuong, J. Kraut, M. Poe, M. Williams, and K. Hoogsteen, Dihydrofolate reductase: X-ray structure of the binary complex with methotrexate. Science 197:452 (1977).

34. R. T. Schimke, P. C. Brown, R. J. Kaufman, M. McGrogan, and D. L. Slate, Chromosomal and extrachromosomal localization of amplified dihydrofolate reductase genes in cultured mammalian cells. Cold Sprg. Harbor Symp. Quant. Biol. Vol. XLV, 785-797 (1980).

TRANSFORMATION AND EXPRESSION OF TK SEQUENCES

Adele El Kareh, Michael Ostrander and Saul Silverstein

Columbia University
College of Physicians and Surgeons
New York, N.Y. 10032

INTRODUCTION

Cells lacking thymidine kinase (Ltk-)can be converted to the tk+ phenotype following exposure to UV-irradiated herpes simplex virus (HSV) and selection in HAT medium (1,2,3,). These experiments suggested to us that isolation of the virus sequences encoding tk was feasible. Accordingly, virus DNA cleaved with a variety of restriction endonucleases was assayed for its ability to convert tk- cells to a HAT resistant phenotype using the calcium phosphate precipitation technique (4). Digestion of HSV DNA with Bam H I resulted in the appearance of numerous colonies following application of the DNA and selection in HAT (5). Subsequent analysis of Bam H I cleaved, size-fractionated DNA identified a unique 3.5 kb fragment that contained the information required to convert tk- cells to the tk+ phenotype. Colonies that arise following application of DNA and HAT selection are said to be transformed. Transformation results from integration of the virus and carrier DNA sequences into cellular DNA (6,7,8,). The cells' transcriptional and translational apparatus recognize the virus tk sequences to form a functional virus specified tk.

The tk gene has been cloned by a number of investigators and the complete nucleotide sequence of two independently derived clones determined (9,10). The direction of transcription and the boundaries of the transcript have been defined (11,12,). These features coupled with the availability of rapid specific assays for the genes' translation product and a sensitive bioassay for gene function makes this an ideal model system for examination of the functional consequences of in vitro

111

manipulation of its sequences. Here we describe experiments to examine the expression and the sequences involved in regulating expression of the tk gene in transformed cells. Our approach was to first define the physical limits required for conversion of tk⁻ cells, these analyses were extended to include a series of sequenced 5' deletion mutants all of which terminated at a common 3' Hind III site (13). We next asked which sequences are required for transactivation of the resident virus tk gene after infection with a tk⁻ HSV.

Previous work demonstrated that superinfection of cells transformed with UV-irradiated virus resulted in an increase in the level of tk (14). This and subsequent studies demonstrated that the increased levels of tk required expression of the superinfecting genome and that ts mutants containing a defect in an immediate early (IE) polypeptide would not transactivate the resident tk when infection proceeded at the nonpermissive temperature (15,16,17). Thus, expression of the gene in transformed cells is regulated in trans by a virus immediate early protein in the same fashion as tk is regulated during the course of productive infection (18). Therefore, we asked which sequences are necessary for transactivation. Here we demonstrate that the sequences required for proficient transformation are indistinguishable from those required for transactivation.

In another study we show that cells which are phenotypically tk⁻ as a result of methylation of CpG pairs within and about the tk sequences can be converted back to the tk⁺ phenotype following growth in 5-azacytidine. These cells now clone in HAT, express tk mRNA and their tk sequences are no longer refractile to digestion with the restriction enzyme Hpa II.

RESULTS

Identification of Sequences Required for Transformation

The identification of a unique restriction endonuclease fragment that could convert tk⁻ cells to the tk⁺ phenotype led us to ask what the boundaries of the gene were. Previous studies in our laboratory and by others (19) suggested that efficient transformation occurred following exposure to DNA bounded by the Pvu II site at nucleotide 575 and the Sma I site at 1910 (Fig.1). The Sma I site is just within the carboxy-terminus of the peptide (10). Subsequent studies have employed fragments derived following cleavage with other restriction endonucleases. In many instances these fragments were subcloned in pBR 322 to insure purity. The results of

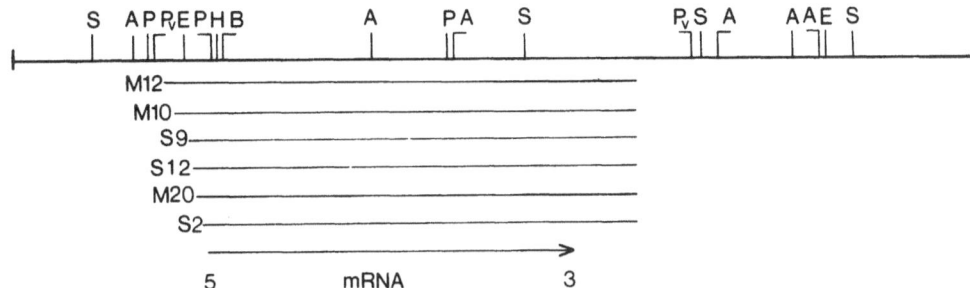

Fig. 1   Map of tk DNA. The letters above the line represent
         naturally occurring restriction endonuclease sites
         within the 3.6 kb Bam H I fragment. Eco R I (E), Pvu II
         (Pv), Hinc II (H), Bgl II (B), Pst I (P), Sma I (S), and
         Alu I (A). The designations below the line refer to the
         deletion mutants (12) used in these studies. They all
         have a unique 3' boundary marked by a synthetic Hind III
         site and varying 5' ends terminated by a synthetic
         Bam H I site.

transformation studies with these fragments are shown in Table
1. The data demonstrate that the 2.0 kb Pvu II fragment is as
efficient as the intact Bam fragment at converting cells to HAT
resistance. We also note that the larger 2.4 kb Eco R I
fragment is relatively inefficient at converting tk⁻ cells.
The map position of these sites suggest that the 5' boundary of
the tk gene lies between the Pvu II and Eco R I sites. The 2.8
kb Bgl/Bam fragment was subcloned and examined for its ability
to convert tk⁻ cells. This fragment was less proficient at
converting tk⁻ cells than the others we tested.

    The cell lines that were derived following transformation
with these various DNAs were examined for the arrangement and
number of integrated sequences by Southern blot hybridization
(20). The results of these analyses demonstrated that lines
derived from DNA whose sequences initiated 3' to the Pvu II site
contained multiple copies of tk DNA, whereas, those derived
following exposure to fragments at or 5' to the Pvu II site
generally contained only a single copy. Further analysis of the
clones generated after exposure to DNA bounded by natural
restriction endonuclease sites demonstrated that sites at or
near the terminii were not maintained. Thus it was difficult to
define the sequences required for proficient transformation.

    To precisely define the nucleotide sequences required for
efficient transformation we utilized a series of ordered 5'
deletion mutants. These mutants, constructed by McKnight,

Table 1. Transformation Efficiency of Restriction Endonuclease
          Fragments of HSV-1 DNA

Fragment	Size (kb)	Clone	Transformation Efficiency[a]	Transactivation
Hpa I	8.4	yes	1.0	yes
Kpn I	5.1	yes	1.0	yes
Bam H I	3.6	yes	1.0	yes
Pvu II	2.0	yes	1.0	yes
Eco R I	2.4	yes	0.05	no
Bgl/Bam	2.8	yes	0.01	no
Hinc/Bam	2.8	no	0.01	ND[b]
Pvu/Sma	1.3	no	0.005	ND

a) transformation efficiency is calculated on a ug equivalent
   basis as described in Wigler et al. (5) and adjusted for
   efficiency relative to the cloned 3.6 kb Bam H I fragment.
b) ND means not done.

maintained a constant 3' end marked by a synthetic Hind III site
and progressively shortened 5' terminii bounded by Bam H I sites
(13). Each of these tk DNAs was cloned in pBR 322 as a unique
Bam/Hind fragment. DNA from clones containing varying lengths
of 5' sequence were digested with Sal I and assayed for their
ability to convert tk⁻ cells. We rationalized that digestion
with Sal I which cuts once in the plasmid DNA sequences would
serve to buffer the ends of the tk gene and thus protect them
from nucleases during the integration process. Table 2 shows
that the efficiency of transformation with the various deletion
mutants ranged from 2700/ug for M12 to 150/ug for S2. For
comparison the transformation efficiency of Sal I cut p TK-5,
p MOE and p Bgl 2.8 were measured. These studies show that
deletions which exceed 109 bp 5' to the mRNA cap site decrease
the transformation efficiency. The Eco R I site at 81
encompasses a highly conserved member of the sequence GGCAAT
found upstream from most structural genes (21). Thus presence
of this sequence appears to be required for proficient
transformation. Mutants 29 bp 5' to this sequence are 25-fold
more efficient in a transformation assay. Deletion mutants
encompassing TATTA at 24 but none of the nucleotides present in
the mRNA are less efficient at transforming cells than those
terminating at the Eco R I site. Deletion past the cap site
results in a further diminution of transforming activity. Thus

Table 2. Functional Analysis of 5' Deletion Mutants

Deletion Mutant	Base Pairs 5' to mRNA	TATTA	GAAT	Transformation Efficiency	Transactivation
S2	6	–	+	0.022	no
M20	31	+	–	0.017	no
S12	51	+	–	0.020	no
M13	56	+	–	0.014	no
S9	66	+	–	0.014	no
M10	109	+	+	0.380	yes
M12	154	+	+	0.400	yes
pTK–5	646	+	+	1.0	yes
pMOE	80	+	±	0.017	no
pBgl 2.8	–52	–	–	0.004	no

a boundary between 81 and 109 bp 5' to the mRNA cap site influences the efficiency of transformation with this gene.

Sequence Complexity of Cells Transformed With Deletion Mutants

Several independent clones were isolated following transformation with each mutant DNA and expanded in HAT medium. The tk sequences present in each line were analyzed by blot hybridization. To determine the size of the integrated tk sequences DNA from each clone was digested with Bam H I and Hind III. The results of this analysis demonstrate two points, i) numerous integration events are associated with the establishment of tk$^+$ cells from deletion mutants S2 and S9 and ii) the majority of 5' and 3' sites are preserved following the integration of mutant DNAs which extend beyond the boundary at 109.

Transactivation of Endogenous Sequences

Lin and Munyon (14) observed that superinfection of cells,transformed to the tk$^+$ phenotype by UV-irradiated HSV, with a tk$^-$ virus resulted in increased levels of virus specified tk. Later studies (16,17) established that transactivation of the endogenous locus required expression of an IE gene. The work of Preston (18,22) has implicated a

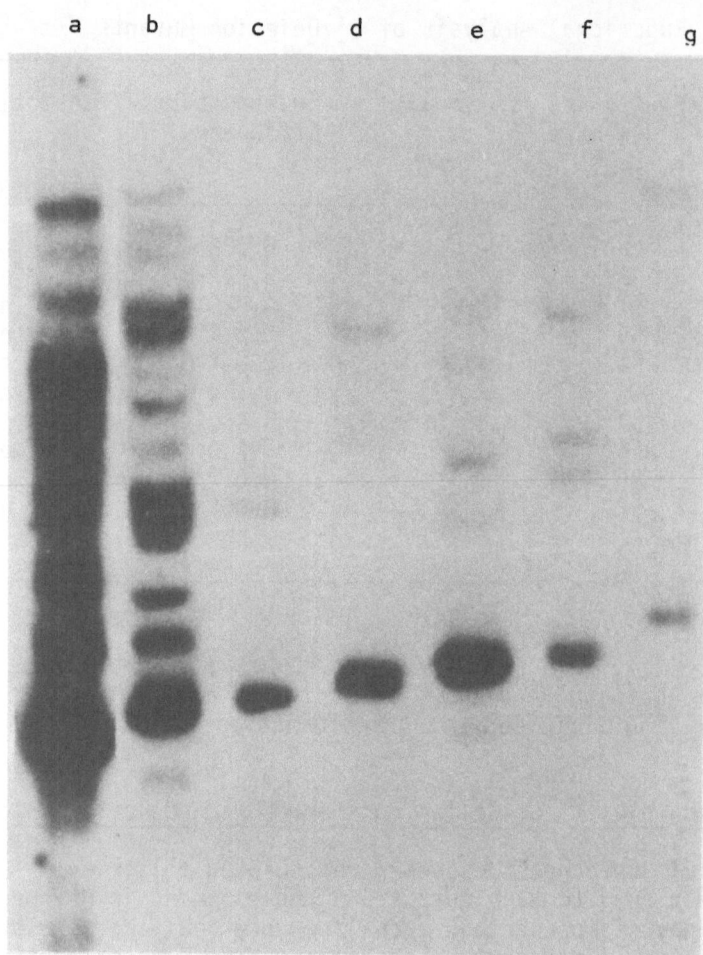

Fig. 2   Blot hybridization profile of transformed cell DNA.
         20 ug of DNA extracted from tk+ transformants derived
         by transfection with deletion mutant DNAs was digested
         with Bam H I and Hind III, electrophoresed on a 0.8 per
         cent agarose gel and blot hybridized using ^{32}P-labeled
         tk DNA as probe. DNAs are from a line derived following
         transformation with S2 (a), S9 (b), M10 (c), M12 (d),
         M14 (e), M11 (f) and M6 (g).

specific virus encoded protein, Vmw 175, in transactivation.
The purpose of the studies which follow was to determine the
specific nucleotide sequences responsive to transactivation.
Initially clones derived by transformation with defined
restriction endonuclease fragments were tested. In these
experiments tk+ or tk- cells containing a single copy of the

virus tk sequences were infected at two different multiplicities
and the activity of tk followed as a function of time
postinfection. The results of a typical study are shown in
Fig.3. We note that the level of tk increases during the course
of infection and that the increase in tk activity is dependent
on the multiplicity of infection. It is noteworthy that as the
m.o.i. exceeds 50 the response decreases. This observation is
expected if expression of late virus functions which turn off
synthesis of proteins such as tk was accelerated because of the
initially high m.o.i. A second interesting observation is that
phenotypically tk$^-$ cells can be reactivated to express tk
following infection with tk$^-$ virus. These analyses were
extended to include tk$^+$ cells derived following transformation
with the 3.5 kb Bam H I fragment. Thus we may conclude that
only the sequences adjacent to the tk gene are required for
transactivation.

To define the boundary responsive to transactivation we
examined the capacity of cells transformed with deletion mutant
or cloned DNAs to increase the level of tk activity in response
to infection with a tk$^-$ virus. Tk$^+$ clones were examined for
the conservation of the appropriate restriction sites at their
5' and 3' boundaries and then infected with tk$^-$ virus. Our
results demonstrate that the sequences between 81 and 109 are
required for transactivation. These are the same sequences that
are required for proficient transformation.

Transcription From Endogenous TK Templates

The size of the tk transcript synthesized in the various
transformants was determined by blot hybridization of poly A
containing RNA using ^{32}P-labeled nick translated tk DNA as
probe. This analysis reveals that deletions extending past 109
[S2, S12, p MOE (not shown)] qualitatively effect the size of
the tk specific transcript in these cells (Fig.4). Mutants
encompassing 109 show no detectable alteration in the size of
their transcripts. We presume that these transcripts lack 5'
sequences because in vitro transcription studies demonstrate
that mutants extending 3' to 109 synthesize shorter transcripts
(13).

We next asked if transactivation resulted in the
accumulation of tk mRNA sequences in the transformed cell.
Transformed cells were infected with a tk$^-$ deletion mutant of
HSV-1. This mutant lacks the 875 bp Pst I fragment found within
the structural sequences of the DNA that codes for tk (22). The
availability of this mutant permits analysis of the "cellular"
tk transcripts without interference from sequences that are
derived from the infecting virus. Accordingly, identical

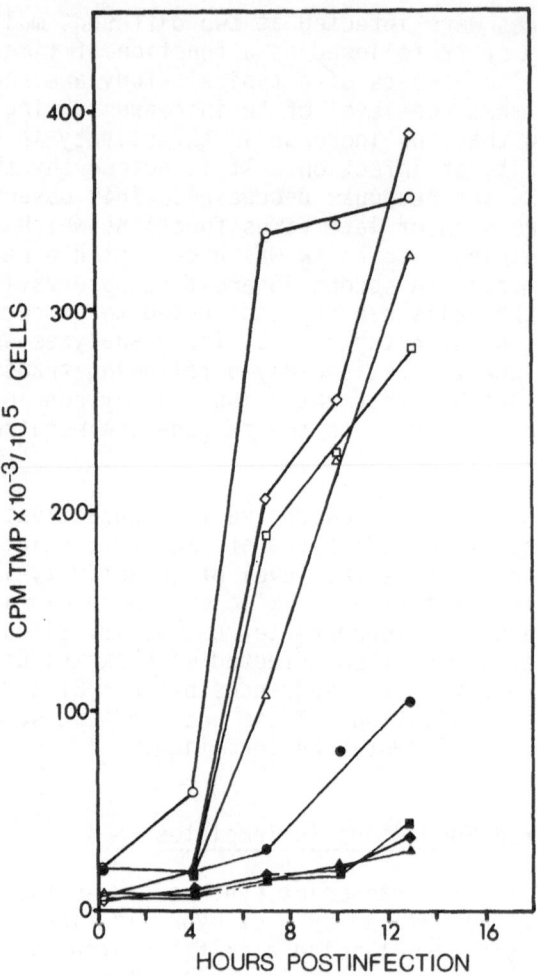

Fig. 3   Transactivation. A tk⁺ cell line LHH5-1○ ○ and
         three phenotypically tk⁻ revertants LHH5-1B1□ □,
         LHH5-1B2◇◇and LHH5-1B3△△were infected at an moi
         of 2 (filled symbols) or 20 (open symbols) with a tk⁻
         mutant of HSV-1 and the level of tk activity measured
         as a function of time postinfection.

amounts of poly A-RNA extracted from infected transformed cells
at intervals postinfection was analyzed by a modification of the
Northern blot hybridization technique. Comparison of the band
intensity obtained following hybridization to RNA isolated from
mock-infected, 6 and 12 hour infected cells demonstrates an
accumulation of tk specific RNA sequences in response to the
signal supplied in trans by the infecting virus (Fig.4).

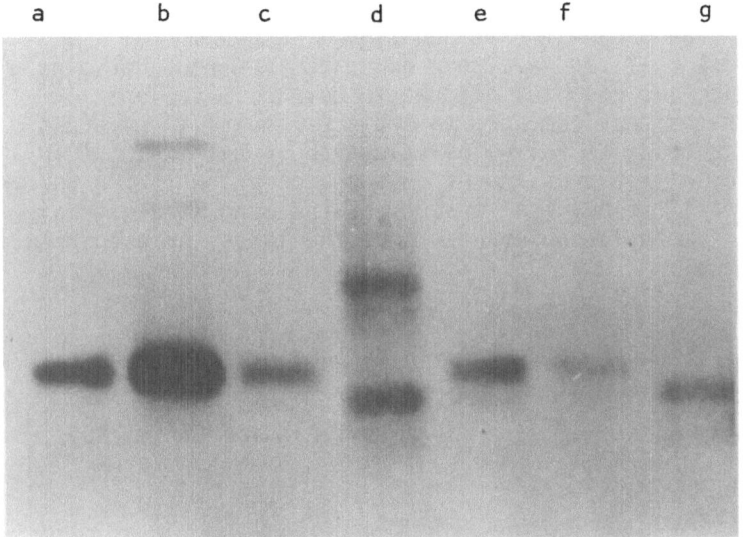

Fig. 4   Synthesis of tk mRNA in transformed and transactivated
         cells. 5 ug of poly A-containing cytoplasmic RNA was
         analyzed by blot hybridization. RNA from a single copy
         Bam H I transformant infected with a tk⁻ deletion
         mutant of HSV-1 (22) for 6 (a) or 12 (b) hours or mock
         infected (c). RNA from cells transformed with deletion
         mutant S9 (d), M10 (e and f) and S12 (g).

Activation of a Non-Reverting Line

     Analysis of the DNA from K_1B_6 ^{me} a non-reverting tk⁻
cell line revealed that the Hpa II sites present within the tk
gene were refractile to digestion by this enzyme. Digestion
with the isoschizimer Msp I revealed a characteristic set of low
molecular weight fragments consistent with a limit digest of the
tk gene. We concluded that the gene was hypermethylated in this
non-reverting variant and that methylation prevents phenotypic
switching (23). On-off transcriptional regulation is associated
with changes in the methylation pattern about genes. Recent
studies have demonstrated that actively transcribed genes are
hypomethylated when compared with their unexpressed counterparts
(24,25). If it is possible to demethylate the locus than it
should be possible to reactivate tk. The cytidine analogue
5-aza C was shown to alter the differentiated state of cultured
mouse embryo cells and to inhibit methylation of newly
synthesized DNA (26,27). Therefore, K_1B_6 ^{me} cells were
exposed to varying concentrations of 5-aza C for 48 hours. The

culture medium was then changed and cells allowed to proliferate in drug free medium for an additional 48 hours. At this time cells were replated in either nonselective, BUdR containing or HAT medium and colonies allowed to develop. Table 3 demonstrates that exposure to 5-aza C results in numerous surviving colonies following selection in HAT medium. The frequency of reexpression of this locus is low (10^{-5}) and this may in part reflect the toxicity of the drug, the extensive number of methylation events about the locus, or a combination of the two.

Table 3. Effects of 5-Azacytidine on a Non-Reverting tk- Line

| | Cloning Efficiency | | |
Drug Concentration	DME	BUdR	HAT
none	0.92	1.18	ND
2 uM	0.68	0.85	1.1×10^{-5}
5 uM	0.33	0.48	7.3×10^{-6}
10 uM	0.25	0.36	2.0×10^{-6}
20 uM	0.30	0.36	1.3×10^{-6}

The methylation pattern about the tk gene was examined by blot hybridization. DNAs from K_1B_6 me and two HAT resistant, 5-aza C reversed clones were digested with Msp I and its methyl-sensitive isoschizimer Hpa II and their annealing profiles were compared. Figure 5 demonstrates that the vast majority of Hpa II refractile sites are no longer present in the reversed clones. Thus we conclude that reversion to the tk$^+$ phenotype is accompanied by a reversal of the modification of this locus in a previously non-reverting cell.

CONCLUSIONS

A series of cloned restriction endonuclease fragments and deletion mutants of the HSV-1 tk gene were used to analyze the role of flanking nucleotide sequences on the efficiency of conversion of Ltk- cells to the tk$^+$ phenotype and the ability of these transformed sequences to be transactivated by

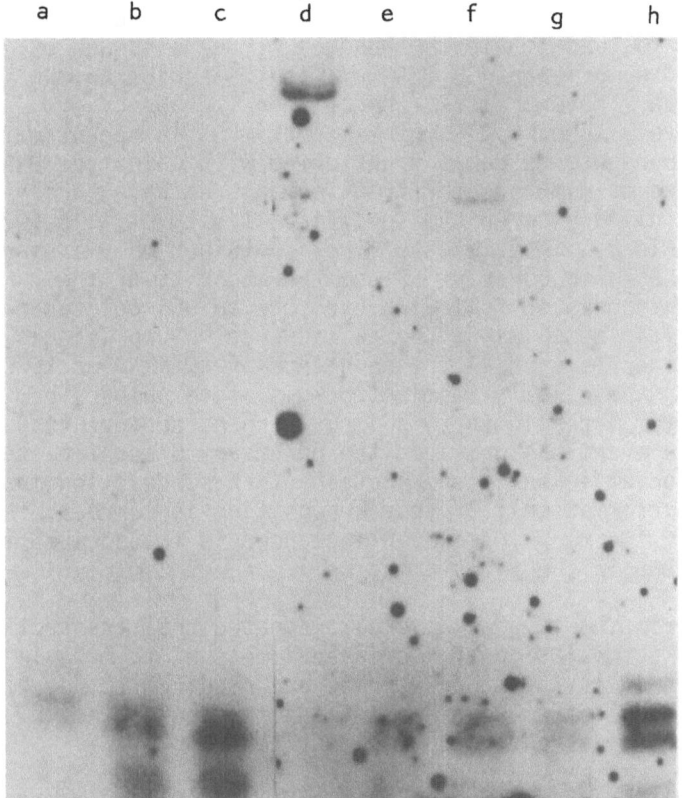

Fig. 5   Blot hybridization profile of tk⁻ cell DNA before and
after growth in 5-aza C. DNA extracted from a non-
reverting tk⁻ cell and two HAT resistant revertants
isolated after growth in 5-aza C was cleaved with Hpa II
(a,b,d,f) or Msp I (c,e,g,h) and blot hybridized using
^{32}P-labeled tk DNA as probe. Lanes a and h, pTK-5,
b and c K_1B_6 aza-1, f and g, K_1B_6 aza-2, and
d and e K_1B_6 ^{me}.

infection with a tk⁻ virus. Our experiments identify a
control region 5' to the mRNA cap site between nucleotides 81
and 109. Deletion past 109 results in a marked decrease in the
efficiency with which tk⁻ cells are converted to the tk⁺
phenotype as well as the loss of the capacity of the transformed
cell to be transactivated. The region identified in this study
is 5' to both the TATTA and CGAAT control sequences found
flanking most eukaryotic mRNAs (21). Thus while these regions
may be important for quantitative and accurate transcription of
the tk gene in vitro they alone do not define all of the
nucleotide sequences required for faithful and efficient
expression of this gene in transformed cells.

Our results permit us to define a role for the sequences at, or just 5' to, CGAAT. In the absence of these sequences, cells transformed with mutant S9 DNA fail to transcribe an mRNA of the correct size. Nevertheless, the cell is tk$^+$ and quantitatively appears to synthesize tk mRNA in approximately the same abundance as cells transformed with wild-type DNA. In our analyses of numerous tk$^+$ transformants we have rarely detected a difference in the specific activity of tk between these cell lines. Most of the lines contained only a single copy of tk DNA, in contrast S9 and the majority of the transformants derived following exposure to DNA deleted past 109 contain multiple copies of the tk sequence. Surprisingly, there is no gross difference in the abundance of tk mRNA or tk enzyme activity in these cells despite the disparate numbers of potential templates in these cells. Perhaps an adventitious integration event has provided the necessary 5' control region required for efficient transcription of these deletion mutants in the transformed cell.If this supposition is correct, than only one of the many integrated sequences is transcribed to yield tk mRNA in the transformed cell.

Transcription studies have demonstrated that transactivation results in accumulation of increased levels of tk mRNA in transformed cells. Further studies are required to demonstrate that this results from increased transcription of the tk locus and not stabilization of the resident tk mRNA.

In preliminary studies we have asked the 5' sequences of the tk gene to serve as a promoter, for a gene from a prokaryote, to determine if these sequences would be recognized and transactivated. The prokaryotic sequences coding for resistance to an aminoglycoside that is effective in both prokaryotes and eukaryotes was cloned into the Bgl I site of tk and used to transform mammalian cells to resistance to G418. Resistant cells were shown to produce a hybrid prokaryote-tk mRNA whose level was increased following infection with HSV. Thus the tk sequences 5' to the Bgl I site can serve to promote transcription of a gene from prokaryotes and these same sequences respond to virus infection by transactivating the endogenous hybrid locus.

Finally, we have provided more evidence of a role for methylation as a means of controlling gene expression in higher eukaryotes. Our studies demonstrate that hypermethylation of the tk gene in transformed cells results in transcriptional inactivation of this locus. When cells containing this hypermethylated locus are grown in 5-aza C they express tk at a low frequency. HAT resistant clones selected after 5-aza C treatment were shown to be hypomethylated. Thus as in other systems (24,25) we have correlated methylation with gene activity and demonstrated transcriptional activation of a quiescent locus after treatment with 5-aza C.

ACKNOWLEDGEMENTS

This research was supported by a grant to SS from the USPHS CA 17477, SS is the recipient of a Research Career Development Award from the USPHS CA 0049. MRO is the recipient of a Postdoctoral Fellowship from the USPHS CA 06543. We thank Steve Vogel for immeasurable help with many of the actual experiments. Dr. J. Smiley of the University of Ontario for making available the tk⁻ deletion mutant of HSV-1 and Dr. S. McKnight of the Carnegie Institute for providing us with the 5' deletion mutants used in these studies.

REFERENCES

1) Munyon, W., Kraiselburd, E., Davis, D. and Mann, R. (1971) Transfer of thymidine kinase to thymidine kinaseless L-cells by infection with ultraviolet-irradiated herpes simplex virus J. Virol., 7, 813

2) Davidson,R.L., Adelstein, S.J. and Oxman, M. (1973) Herpes simplex virus as a source of thymidine kinase for thymidine kinase deficient mouse cells: Supression and reactivation of the viral enzyme. Proc. Natl. Acad. Sci. U.S.A., 70, 1912

3) Munyon, W., Buchsbaum, R., Paoletti, E., Mann, J., Kraiselburd, E. and Davis, D. (1972) Electrophoresis of thymidine kinase activity synthesized by cells transformed by herpes simplex virus. Virology, 49, 683

4) Graham, F.L. and van der Eb, A.I. (1973) A new technique for the assay of infectivity of human adenovirus 5 DNA. Virology, 52, 456

5) Wigler, M., Silverstein, S., Lee, L.S., Pellicer, A., Cheng, Y.C. and Axel, R. (1977) Transfer of purified herpes virus thymidine kinase gene to cultured mouse cells. Cell, 11, 223

6) Pellicer, A., Wigler, M., Axel, R. and Silverstein, S. (1978) The transfer and stable integration of the HSV thymidine kinase gene into mouse cells. Cell, 14, 133

7) Robins, D.M., Ripley, S., Henderson, A.S. and Axel, R. (1981) Transforming DNA integrates into the host chromosome. Cell, 23, 29

8) Perucho, M., Hanahan, D. and Wigler, M. (1980) Genetic and physical linkage of exogenous sequences in transformed cells. Cell, 21, 309

9) McKnight, S.L. (1980) The nucleotide sequence and transcript map of the herpes simplex virus thymidine kinase gene. Nucl. Acids Res., 8, 5949

10) Wagner, M.J., Sharp, J.A. and Summers, W.C. (1981) Nucleotide sequence of the thymidine kinase gene of herpes simplex virus type-1. Proc. Natl. Acad. Sci. U.S.A., 78, 1441

11) Smiley, J.R., Wagner, M.J., Summers, W.P. and Summers, W.C. (1980) Genetic and physical evidence for the polarity of the thymidine kinase gene of herpes simplex virus. Virology, 102, 83

12) McKnight, S.L. and Gavis, E.R. (1980) Expression of the herpes thymidine kinase gene in Xenopus laevis oocytes: An assay for the study of deletion mutants. Nucl. Acids Res., 8, 5931

13) McKnight, S.L., Gavis, E.R., Kingsbury, R. and Axel, R. Analysis of the HSV thymidine kinase gene: Identification of an upstream control region. (submitted for publication)

14) Lin, S.S. and Munyon, W. (1974) Expression of the viral thymidine kinase gene in herpes simplex virus transformed L-cells. J. Virol., 14, 1199

15) Garfinkle, B. and McAuslan, B.R. (1974) Regulation of herpes simplex virus induced thymidine kinase. Biochem. Biophys. Res. Commun., 58, 822

16) Leiden, J.M., Buttyan, R. and Spear, P.G. (1976) Herpes simplex virus gene expression in transformed cells 1: Regulation of the viral thymidine kinase gene in transformed L cells by products of superinfecting virus. J. Virol., 20, 413

17) Kit, S., Dubbs, D.R. and Schaffer, P.A. (1978) Thymidine kinase activity of biochemically transformed mouse cells after superinfection by thymidine kinase negative temperature sensitive herpes simplex virus mutants. Virology, 85, 456

18) Preston, C.M. (1979) Control of herpes simplex virus type 1 mRNA synthesis in cells infected with wild-type virus or the temperature sensitive mutant tsK. J. Virol., 29, 275

19) Colbere-Garapin, F., Chousterman, S., Horodniceanu, F., Kourilsky, P. and Garapin, A. (1979) Cloning of the active thymidine kinase gene of herpes simplex virus type 1 in Escherichia coli K-12. Proc. Natl. Acad. Sci. U.S.A., 76, 3755

20) Southern, E.M. (1975) Detection of specific sequences among DNA fragments separated by gel electrophoresis. J. Mol. Biol., 98, 503

21) Efstradiatis, A., Posakony, J.W., Maniatis, T., Lawn, R.M., O'Connell, C., Spritz, R.A., DeRiel, J., Forget, B.G., Weissman, S.A., Slightom, J.L., Blechl, A.E., Smithies, O., Baralle, F.E., Shaulders, C.C. and Proudfoot, N.J. (1980) The structure and evolution of the human β-globin gene family. Cell, 21, 653

22) Smiley, J. (1980) Construction in vitro and rescue of a thymidine kinase deficient deletion mutant of herpes simplex virus. Nature, 285, 333

23) Sweet, R., Jackson, J., Lowy, I., Ostrander, M., Pellicer, A., Roberts, J., Robins, D., Sim, G.-K., Wold, B., Axel, R. and Silverstein, S. (1981) The expression, arrangement and rearrangement of genes in DNA-transformed cells, in Genes, Chromosomes and Neoplasia, F.E. Arrighi, P.N. Rao, and E. Stubblefield, ed. Raven Press, N.Y.

24) McGhee, J. and Ginder, M. (1979) Specific DNA methylation sites in the vicinity of the chicken β-globin genes. Nature, 280, 419

25) Weintraub, H. and Groudine, M. (1981) α-Globin gene switching during the development of chicken embryos: Expression and chromosome structure. Cell, in press

26) Taylor, S.M. and Jones, P.A. (1979) Multiple new phenotypes induced in 10T1/2 and 3T3 cells treated with 5-azacytidine. Cell, 17, 771

27) Jones, P.A. and Taylor, S.M. (1980) Cellular differentiation, cytidine analogs and DNA methylation. Cell, 20, 85

28) Rothstein, S.J., Jorgensen, R.A., Postle,. K. and Reznikoff, W.S. (1980) The inverted repeats of Tn5 are functionally different. Cell, 19, 795

GENETIC VARIANTS OF CULTURED ANIMAL CELLS

Peter N. Ray and Louis Siminovitch

Department of Genetics
Hospital for Sick Children
Toronto, Ontario, Canada

INTRODUCTION

It has been clear for a number of years that the functional analysis of biological systems is highly dependent upon the availability of a large number of mutants of broad phenotypic classes. It was with this understanding that our laboratory undertook about ten years ago to develop methods for selection of mutants in somatic cells. Over this decade the field has developed extremely rapidly both in terms of methodological expertise and in the wide spectrum of mutants which can now be used for studies on gene function and regulation.

In this review we shall present an overview of the mutational systems available in somatic cells. Although the survey will attempt to be as exhaustive as possible, we cannot claim that it will be all-inclusive. Our general intention is to emphasize some generalities about the state of the field of somatic cell genetics using some of the systems which have been described as prototypes. We shall restrict our discussion to mutants selected in vitro and thus not include the extensive number of cell strains derived from patients with genetic diseases.

Several classes of mutants have been described[1,2,3]. These include conditional lethal isolates of the temperature sensitive or auxotrophic kind, mutants resistant to various drugs, and mutants involving defects in specific biological function such as the immunoglobulins[4].

Conditional lethal mutants have nearly always been isolated by indirect selection methods, akin to procedures developed earlier in microorganisms[5,6,7]. However, in several cases, the isolates have been obtained inadvertantly after selection for some other phenotype.

CONDITIONAL LETHAL MUTANTS - AUXOTROPHS

A listing of the auxotrophic mutants is shown in Table 1. These systems have been the subject of a great deal of work mainly by Puck, Kao, Patterson and their collaborators using CHO cells[5,6,11,12] and we only wish to highlight some general features of the systems.

(1) Because of the extensive nutritional requirement of cells grown in culture, only a limited spectrum of mutations to auxotrophy are available for isolation. Nevertheless the list in Table 1 represents a very large fraction of the potential auxotrophic mutants. Most of the mutants listed have been obtained after selection. However this is not always the case. The nutritional requirements of cell lines developed for growth in culture (unselected) can vary from one line to another. For example, very early in their studies Puck and his associates found that CHO-K1 cells were naturally auxotrophic for proline[10]. Some cell lines cannot use cystathionine in place of l-cystine, citulline instead of l-arginine, and branched chain α-keto acids instead of the corresponding l-amino acids (e.g. isoleucine, leucine, valine)[42,43]. Thus there seems to be specific enzyme deficiencies in the tissues of animals from which these cell lines originated.

(2) All the auxotrophs behave recessively. This finding, combined with (1) above, and the fact that we are dealing with near diploid cells, represents a rather remarkable situation which will be discussed later.

(3) The exact biochemical lesion has been identified for a large majority of the mutants. For example the nine adenine requiring auxotrophs represent mutations in eleven of the twelve biochemical steps involved in the synthesis of AMP from phosphoribosyl pyrophosphate[11,19].

This is also true for the two classes of mutants which are auxotrophic for glycine, thymidine, and purines. In the first one shown in Table 1 the lesion involves a defect in folylpolyglutamate synthetase, an enzyme which adds glutamate residues to folic acid[20-22]. The second type of isolate obtained in Chasin's laboratory by ^{3}H-deoxyuridine suicide[23] involves a defect in dihydrofolate reductase; as will be pointed out later, the latter mutants represent one of 3 types of mutation involving this target enzyme.

(4) The uridine requiring auxotrophs provide insight into the genetic organization of mammalian cells. The three complementation groups together show defects in all six of the enzymatic reactions

Table 1. Auxotrophic Mutations

Phenotype	Number of Complementation Groups	References
Glycine	4	6,8,9
Proline	1	10
Adenine	9	8,11-19
Glycine, adenine, thymidine	1	11,12,20-22
Glycine, hypoxanthine, thymidine (^{3}H deoxyuridine resistant)	1	23
Adenine, thymidine	1	11,12
Thymidine (AraC^r)	1	24
Thymidine	1	25
Uridine	3	26-31
Sterols	2-3	32-36
Serine	1	37
Glutamate	1	38
Alanine	1	38
Inositol	1	6
Asparagine	1	39
CO_2, asparagine (Respiration Deficient)	7-8	40,41

required for the biosynthesis of UMP[26-31]. One class of mutants called uridine A has lost the first three activities, carbamyl
phosphate synthetase, aspartate transcarbamylase, and dihydro-
orotase. The coordinate loss of these three activities was shown
to be due to a single mutational event in that revertants of the
mutant simultaneously regained all three activities[26,27]. This
conclusion was confirmed by the purification of a single polypeptide having all three enzymatic activities[27,44]. In addition,
as will be discussed later, the three enzymatic activities are
coordinately increased in mutants selected for resistance to a

drug which inhibits one of the activities, aspartate transcarba-mylase[45]. Parts of the gene coding for these activities have subsequently been cloned[46].

A similar example of a single gene coding for a multifunctional enzyme is seen in another class of uridine auxotrophs called uridine C. This mutant was isolated as being resistant to the drug 5-fluroorotic acid and was found to have lost the last two enzymatic activities in UMP biosynthesis, orotate phosphoribosyl transferase, and orotidylate decarboxylase[29,31]. Thus, in the UMP biosynthetic pathway five of the six enzymatic functions are controlled by two genes.

(5) The availability of the auxotrophic systems offers several important advantages for studies in somatic cells and molecular genetics.
(a) By their nature, revertants can be obtained by relatively simple selection procedures. In fact, in at least three cases, revertants have been isolated with a ts phenotype providing convincing evidence that the auxotrophy is due to a structural gene mutation[22,39,47]. In addition, the reversion systems provide excellent material for studies of mutagenesis.
(b) The lesions can and have been used for selection of somatic cell hybrids.
(c) The availability of comprehensive sets of mutants covering most if not all of the biochemical steps in a pathway make it possible to initiate a more complex genetic analysis of the regulation of the pathway. This is currently being done with the adenine auxotrophs[48,49].
(d) Because the wild type allele is dominant in conditional lethal systems, the mutants themselves can be used as recipients in gene transfer experiments. This should make it possible to map these functions and eventually clone the genes involved.

CONDITIONAL LETHAL MUTANTS - TEMPERATURE SENSITIVE

Temperature sensitive mutants, either heat or cold sensitive, represent the second class of conditional lethal isolates (Table 2). The general technology used in selecting such mutants is similar to that employed for obtaining auxotrophs[7,50-53]. The following represents some of the interesting features of these mutational systems.
1) Because of the inherent nature of ts mutations, it is not surprising that the spectrum of phenotypes observed has been rather wide. These include CHO mutants with altered aminoacyl-tRNA synthetases[54-65], BHK mutants with altered 28S ribosomal RNA production and defective RNA synthesis[7,68-72], and a variety of mutants blocked at specific stages of the cell cycle[52,53,73-81]. Although it is known that the latter cells are defective in their ability to traverse the cell cycle, in most instances the molecular defect in the cell cycle mutants has not

Table 2. Temperature Sensitive Mutations

Phenotype	Number of Complementation Groups	References
Aminoacyl-tRNA synthetase	8	53-65
28S ribosomal RNA synthesis	1	66, 67
DNA synthesis	2-3	68-72
Cell cycle and undefined	ca 10	73-83
Specific enzymes		
hypoxanthine phosphoribosyl transferase	1	84,85
RNA polymerase II	1	86
thymidine kinase	1	87

been ascertained. In several other cases conditional lethal ts
mutants have been obtained in which the lesions have not been characterized in any way[78].

2) All the ts mutants behave recessively. Thus this collection of
isolates (at least 30 mutants) provides another illustration of the
paradox mentioned in respect to the auxotrophs; that is, how does
one account for the great success in obtaining several types of
recessive mutants in quasi-diploid cells? As with the auxotrophs,
in at least one case (the aminoacyl-tRNA synthetases), it has in
fact been possible to obtain a large fraction of the potentially
isolatable mutants[54-65].

3) Since it is very probable that ts mutations involve a protein
which is rendered temperature sensitive because of the presence of
a "faulty" amino acid, there is good reason to believe that all the
isolates in this class represent structural gene mutations. In
some cases, such as the aminoacyl-tRNA synthetases, it has been
possible to demonstrate a temperature sensitive protein[60,63].

4) In some cases selection protocols have been designed to obtain
specific types of ts lesions. Based on the properties of a specific mutant with a ts lesion in leucyl-tRNA synthetase, Thompson and
his colleagues were able to develop procedures using ^{3}H labelled
amino acids as the killing agents to isolate a series of ts mutants
altered in many of the aminoacyl-tRNA synthetases[55,59,61].
Meiss et al. and Roufa have selected cells specifically defective
in DNA synthesis by adding the killing agent to S phase
cells[72,78]. In the latter situation it appears that the defective genes lie on the X chromosome[81]. Methods have been described for obtaining hypoxanthine phosphoribosyl transferase temperature sensitive isolates[84,85], RNA polymerases II mutants
which are ts for growth[86], and mutants with a thermolabile
thymidine kinase[87]. Some investigators have succeeded in selecting cells which are temperature sensitive in their ability to
express a tumor phenotype, i.e. the capacity to form colonies in
soft agar[88].

5) Although most of the work in this area has been done with isolates which are heat sensitive, the reciprocal type, i.e. cold
sensitive mutants, have also been obtained. These include mutants
which are affected in their cell cycle[89-91], in tumor phenotype[88], and in one case, mutants in which inhibition of growth
at lower temperature is associated with colchicine resistance[92].

    The ts and cs mutants have uses similar to those described
earlier for the auxotrophs. Since revertants are obtained easily,
the original isolates can be used for mutagenesis work. As before,
since the wild type allele acts dominantly, they should serve as
excellent recipients in gene transfer studies.

DRUG RESISTANT MUTANTS

   A third and probably the most varied class of mutants which has
been developed in somatic cells involves resistance to drugs.
Since the list of such mutants is very long, it will be impossible
to outline the properties of every system, even in a general
sense. However, in the following tables we have attempted some
sort of grouping and plan to single out some particular systems in
order to make specific points which we believe are of interest.

   Before addressing the information in the tables on drug resistance, the following generalizations can be made.
1) Although not actually tested in every system, a large fraction
of the mutants obtained behave recessively.
2) In several cases systems are available for counter selection of
revertants.
3) Drug resistance is often directed at a specific target. But as
is shown in the tables, in several cases, the resulting phenotype
may be different from the target, and in some cases, more than one
phenotype results from selection with one drug. Conversely one
particular phenotype is often derived from different selections.
4) In many cases, usually for the above reason, the mutant can be
described in a general sense but the lesion has not yet been defined at the molecular level.

   With these generalizations in mind, we now propose to list and
discuss some of the systems in greater detail. The mutants involved in nucleic acid metabolism are grouped in tables 3 and 4.

   Purine nucleotide phosphorylase mutants have been obtained in
two laboratories and are of interest because they should serve as a
model for a specific known genetic disease[94,95]. An association has been found in man between a deficiency in this enzyme and
severe deficiency of the T lymphocytes of the immune system[96].

   The ara C and ara A resistant mutants represent typical examples in which one selection procedure can result in mutants with
several different lesions. The converse is also illustrated in
this table since both adenosine kinase and ribonucleotide reductase
mutants can be obtained from selections with several different
drugs. In the case of ara C, the resulting thymidine auxotrophs
found by Meuth have increased mutation frequencies for some markers[24,101]. Actually the basic biochemical lesion in these
mutants has not yet been identified.

   There are several points of interest about the methotrexate
resistant mutants. Two classes of mutants are shown here because
only these affect the enzyme dihydrofolate reductase[103-111].
There is another class which is found in such selections which
involves transport of the drug (see later). In this system,

Table 3.   Drug Resistant Mutations

Selection	Lesion	Remarks	References
Resistance to:			
5-bromodeoxycy-tidine	Deoxycytidine deaminase	Revertant selection possible	93
6-thioguanosine	Purine nucleo-tide Phosphorylase	Model for human genetic disease	94,95
Ara C	Deoxycytidine kinase		24,97-100
	Ribonucleotide reductase		
	Thymidine auxotrophy	Increased mutation frequency	
Ara A	Adenosine kinase		102
	Ribonucleotide reductase		
	5-adenosyl homo-cysteine hydro-lase	Increased mutation frequency	
Methotrexate	Dihydrofolate reductase (structure)	Dominant	103-111
	Increased acti-vity of dihydro-folate reducatse	Dominant - one class stable & derived from structural mutant - second class unstable - often involves double minute - gene amplifi-cation	
2-deoxyglucose	Galactokinase		114
Hydroxyurea	Ribonucleotide reductase	Dominant - structure gene amplification	115-119
Aphidicolin	Ribonucleotide reductase	Gene amplification - UV sensitivity - hypermutable	120,121

structural mutants can be obtained in single step selections and these behave dominantly[103,104]. The second major class of mutants includes cells with increased levels of the target enzyme[105-111]. As was mentioned earlier in connection with the auxotrophs, work in Chasin's laboratory has shown that one can also isolate mutants which lack dihydrofolate reductase activity. Thus one can obtain mutants missing the enzyme, mutants with structural gene changes in the enzyme, and mutants in which the gene coding for the enzyme is amplified.

These gene amplification mutants are of interest for several reasons. They represent the most thoroughly studied of a common class of mutants where resistance to a drug results from amplification of the gene. The amplified lines have variable degrees of genetic stability. In one case, where the mutants have been obtained as second step isolates of structural gene CHO mutants, the lines contain about 10 fold increased activities of the altered enzyme and are completely stable[103,104,106]. In other cases where the mutants have been obtained by growth in increasing concentrations of methotrexate, the less stable lines contain large numbers of double minute chromosomes[108], and the more stable ones show an expanded homogeneously staining region[108,111].

Because the double mutant involving the structural gene alteration is stable, and acts dominantly, it has been used extensively as a donor for gene transfer work[112,113].

Resistance to hydroxyurea shows similar characteristics to the methotrexate system in that mutants involving both structural and amplified changes in the target enzyme, ribonucleotide reductase, have been observed. The isolates behave dominantly[115-119], and the gene can be transferred by means of DNA[112]. However, this system has not been studied as thoroughly as the methotrexate system and, other than increased enzyme activity, no molecular evidence has been obtained for gene amplification. Ribonucleotide reductase mutants can be obtained using several different selection methods (see Table 3) perhaps because of the central role of the enzyme in nucleic acid metabolism.

Other mutation systems involving nucleic acid metabolism are shown in Table 4. There are several interesting aspects of this table as well. The toyocamycin and tubericidin selections represent just one way of obtaining adenosine kinase mutants. But more interesting perhaps is the fact that one can observe variations in frequencies of 10^{-3} to 10^{-7} for this locus in different lines and similarly large differences in different sublines of CHO cells[122-125]. The 10^{-3} frequency (without mutagenesis) is one of the highest that has been observed in mammalian cells, and may indicate that these alterations are not due to structural gene mutations.

Table 4. Drug Resistant Mutations

Selection	Lesion	Remarks	References

Resistance to:

Selection	Lesion	Remarks	References
Adenosine, toyo-camycin tubericidin, 2-fluoroadenosine	Adenosine kinase	Recessive - very high mutation frequency in some lines ˖ two way selection	122-125
8-azaguanine, 6-thioguanine	Hypoxanthine phosphoribosyl transferase	Recessive - two way selection	126-130
8-azaadenine, 6-mercaptopurine	Adenine phospho-ribosyl transferase	Recessive - two way selection	131-133
5-Bromodeoxyuri-dine	Thymidine kinase Unknown lesions	Recessive - two way selection	134-138
^{3}H uridine plus fluorouridine	Uridine kinase	Recessive - two way selection	139,140
N-(phosphonacetyl) -L-aspartate	Carbamyl phosphate synthetase, Aspartate transcar-bamylase, Dihydroorotase	Amplified gene system	45,141
5-Fluoroorotic acid, pyrazofurin 6-azauridine, 5-fluorouracil	Orotate phosphori-bosyl transferase Orotidylate decar-boxylase	Amplifed gene system	29-31,142

The hypoxanthine phosphoribosyl transferase[126-130], adenine phosphoribosyl transferase[131-133], and thymidine ki-nase[134-136] systems have been used extensively in somatic cells. Since these lesions as well as the adenosine ki-nase[122-125] and uridine kinase lesions[139,140] affect nucleotide salvage pathways, mutants of this type necessarily employ the endogeneous pathways for nucleic acid biosynthesis. Therefore the use of drugs which inhibit the endogenous pathways allows for the selection of revertants with intact scavenger pathways. The availability of these two way selection systems has been useful not only for obtaining revertants, but in hybrid isolation, and in selections for gene transfer. Three of these

gene systems have been successfully transferred by means of
DNA[112,143-151]. The HPRT system has been particularly useful
in somatic cell genetics because the gene is situated on the X
chromosome, and it is therefore present in only one functional copy
in all diploid or quasidiploid cells[152,153].

Stark and his colleagues have convincingly demonstrated that
resistance to the drug N-(Phosphonacetyl)-L-aspartate is due to
amplification of the gene coding for the first three enzyme activities of the pyrimidine pathway[45,141]. There is also good evidence that resistance to the drugs pyrazofurin, and 6-azauridine is
due to simultaneous increases in the last two enzymes of the pyrimidine pathway[30,142]. As was indicated earlier uridine auxotrophs exist which show loss of these activities, providing other
examples of mutant systems in which an enzyme activity can be
either amplified or lost.

The N-(Phosphoracetyl)-L-aspartate and the methotrexate resistance systems are the best studied examples of gene amplification.
Other systems which may involve gene amplification are listed in
Table 5. Studies on these systems may provide some insight into
the genesis of multi-gene families[162].

A spectrum of mutants which seem to act at the membrane level
is summarized in Table 6. It is obvious that there are already

Table 5. Gene Amplification Systems

Gene Product Overproduced	References
Dihydrofolate reductase	107
Ribonucleotide reductase	99,118
Carbamyl phosphate synthetase, aspartate transcarbamylase, dihydrooratase	45,141
Orotate phosphoribosyl transferase, orotidylate decarboxylase	142
P-glycoprotein in colchicine resistant cells	154-156
Asparagine synthetase	157,158
Ornithine decarboxylase	159
Adenylate deaminase	160
Hydroxymethylglutaryl CoA reductase	161

Table 6. Drug Resistant Mutations

Selection	Lesion	Remarks	References
Resistance to:			
Lectins	Membrane glyco-proteins & glycolipids	Most recessive, 9-10 complementa-tion groups	163-173
Tunicamycin	Membrane glyco-lipids?	Dominant	174
Methotrexate	Transport	Recessive - revertant selec-tion possible	103,104
5-Fluorotryptophan	Transport		175,176
L-Phenylalanine	Transport		177
^{3}H amino acid suicide	Amino acid trans-port		178
Chromate	Sulfate transport	Recessive	179
Methylglyocal bis guanyl hydrazone	Polyamine trans-port		180
Colchicine	Membrane glyco-protein	Dominant - probable gene am-plification - pleiotropic	154-156 181-183
Ouabain	Na^+/K^+ ATPase	Dominant	184
Glucocorticoids	Steroid receptors		185-187
Diphtheria toxin	Membrane		188
Specific antisera	HL-A alterations		189

several different types of mutants which fall into this classifica-
tion. The lectin resistant mutants are particularly interesting
because work by Stanely, Kornfeld and others have shown that several different classes can be obtained, all affecting specific
facets of membrane glycoproteins and glycolipids[163-173]. Most
of these behave recessively, and most are obtained with relatively
high mutation frequencies[168,170]. Together with the conditional lethal and several other drug resistant systems, they illustrate again the relative ease with which recessive mutants can be
obtained in somatic cells. The mutants resistant to the lectin
wheat germ agglutinin are of considerable interest because the
specific membrane alterations these cells possess have been correlated with the metastatic potential of tumor cells[190-192].

Although there are other interesting mutant systems in this
table, we will restrict our remarks to the colchicine system,
studied most extensively by Ling and his group[154-156,181-183].
Mutants resistant to this drug show increased amounts of a 175K
protein on their cell surface, probably due to amplification of the
gene[154-156]. The lesion results in decreased uptake of the
drug. Of particular interest is the fact that this decreased uptake lesion is relatively unspecific and the uptake of several
other unrelated drugs is affected at the same time[182]. The
result is that such mutants show pleiotropic resistance involving a
wide spectrum of toxic substances. Many of these are used in chemotherapy and thus it is important to be aware of this situation in
combined drug treatment.

PROTEIN SYNTHESIS MUTANTS

Table 7 groups mutant systems involving defects in some aspects
of protein synthesis. There are just a few brief comments to be
made here.
(1) The emetine system is of interest for a few reasons. First,
the mutants act recessively and 3 complementation groups have been
described. Secondly, since such mutants seem to occur with relatively high frequency in some lines and are always stable they
serve as an excellent system for mutagenesis studies (see later).
(2) Resistance to diphtheria toxin represents another illustration
of the development of several phenotypes after selection with one
drug. In this case two major phenotypes are observed, one involving an alteration in the protein synthetic machinery and the other
affecting transport of the drug across the cell membrane (see Table
6). In addition, there seem to be subclasses of mutants within the
protein synthesis category[202]. Since some of the mutants act
dominantly and can be readily selected this marker has been used in
studies on mutagenesis in human cells[210,211].
(3) Resistance to α-amanitin was one of the earliest structural
gene mutations observed. Ingles has developed specific selection
protocols to obtain ts mutations of this enzyme[86] and he has

Table 7. Drug Resistant Mutations

Selection	Lesion	Remarks	References
Emetine	40S ribosome	Recessive 3 complementation groups	193-198
Trichodermin	60S ribosome	Recessive	199
Diphtheria toxin	Protein synthesis (EF2)	Probably more than one class	188,100-202
α-amanatin	RNA polymerase II	Dominant-ts mutants	86,203-208
Pactomycin	Morphology	Recessive	209
Aspartyl hydroxamate	Increased levels of asparagine synthetase	Gene amplification	157,158

also recently shown that the gene can be transferred by means of DNA (personal communication).

Table 8 is essentially a pot-pourri of other drug resistance mutants and again we will only make a few brief comments.

The cAMP mutants are of particular interest since there seem to be at least 5 classes altered in their cAMP systems, and one of these may be a regulatory defect[222]. As shown in the table the mutants include:
(1) Cells defective in β-adrenergic receptor function, but with normal receptivity to other modulators of adenyl cyclase activity[214].
(2) Mutants defective in the interaction between hormone binding and activation of the adenylate cyclase[215].
(3) Isolates lacking adenylate cyclase activity[216,217].
(4) Mutants in which the cAMP-dependent protein kinase is defective[218-222]. There seem to be several classes of such isolates. These can involve the catalytic or regulatory subunit of the enzyme. The mutants can have either an increased and/or an absent response to cAMP. The former seems to act dominantly and the latter recessively[222]. In one case in which the kinase is unresponsive to cAMP, evidence has been presented that the cells carry a transdominant mutation which affects the expression of the enzyme[221]. This is one of the few examples of an apparent regulatory mutation in somatic cells.

Table 8. Drug Resistant Mutations

Selection	Lesion	Remarks	References
Canavanine	Argininosuccinate synthetase		212
25-OH cholesterol	Regulation of cholesterol synthesis		161
Sindbis virus	Translational control (?)	Recessive	213
cAMP	Hormone receptor		214
	Interaction between binding & adenylate cyclase activation		215
	Altered adenylate cyclase		216,217
	cAMP-dependent protein kinase	Dominant & recessive - Regulatory & Catalytic subunit One regulatory mutant	218-222
	Deathless		223
Ricin	Unknown	Dominant	168,170
Colcemid	Microtubule protein	Dominant	224-226
α-Methylornithine (α-Difluoro-methylornithine)	Ornithine Decarboxylase	Altered Enzyme Gene Amplification	159
Oligomycin, rutamycin venturicidin, antimycin	glycolysis	Recessive	227
Antimycin, rutamycin chloramphenicol	Mitochondrial		228-231

(5) The fifth kind of isolate has been called deathless because cells do not grow but survive in the presence of cAMP. However there are no alterations in levels of activity or responsiveness to effectors for any of the enzymes in the cAMP system[223].

The ricin mutation is of interest because it is the only example of a dominant behaving lectin resistant isolate. The basis of the lesion is not known.

The original colchicine resistant mutants were isolated in order to obtain cells altered in their microtubule proteins and, as discussed earlier, those attempts resulted only in membrane mutants. However more recently workers in Ling's and Gottesman's laboratories have succeeded in obtaining isolates which do show alterations in their microtubule proteins using different selection procedures involving the drug colcemid[224-226].

The table shows another example of a system which may involve amplification. Work from Scheffler's lab has shown that mutants resistant to α-methylornithine have increased activity of ornithine decarboxylase and preliminary evidence has suggested that this is due to gene amplification[159].

Finally, as the data in the table indicate, several investigators have succeeded in obtaining mitochondrial mutations. The mitochondrial site has been demonstrated by showing that the lesion is cytoplasmically inherited in fusions between enucleated mutant cells and nucleated wild type cells[228-231].

INDIRECT SELECTION METHODS

In many cases mutants have been obtained (Table 9) by indirect selection methods. These have involved sib selection, such as that used by Chasin to obtain glucose-6-phosphate dehydrogenase defective isolates[232], and replica plating which has yielded α-mannosidase negative[233] and UV sensitive mutants[237-239].
Thompson and his group developed a specific technique in which they first treated cells with low doses of an agent such as UV or mitomycin, and picked the colonies which showed impaired growth[238]. In this way they isolated mutants with increased sensitivity to UV, mitomycin and ethyl methane sulfonate[238].
It is of interest that 6 complementation groups for UV sensitivity have been identified among isolates obtained using the latter method, and by replica plating. Obviously mutants of this class and those involving EMS and mitomycin sensitivity provide very useful material for studies of repair processes. The α-mannosidase mutants are of interest because the defect is typical of cells from patients with genetic diseases, although the lesions may not be the same. Such methods may therefore be very promising for obtaining lines which mimic genetic diseases and which would then

Table 9. Mutants Isolated by Sib Selection
or Replica Plating Methods

	References
Glucose-6-phosphate dehydrogenase	232
α-mannosidase	233
Dihydroorotate dehydrogenase	234
Myoinositol auxotroph	235
Phosphatidylcholine synthesis	236
UV sensitivity (6 complementation groups)	237-239
Mitomycin sensitivity	238
Ethylmethane sulfonate sensitivity	238
Immunoglobulin synthesis	4

be much easier to work with than diploid fibroblasts. These techniques have not been exploited extensively and it is probable that other potentially interesting mutants could be isolated in a similar manner.

A variant of an indirect selection method has been used to obtain mutant cells defective in immunoglobulin synthesis, assembly or secretion. In this case mutagenized immunoglobulin producing myeloma cells were plated in soft agar and variants were detected by their inability to produce immune precipitates around the colonies when overlaid with specific antisera.

DISCUSSION

The above discussion makes it quite evident that several classes of mutant systems are now available in somatic cells. We have tried to make some generalizations as we went along but will reinforce these now.

(1) The point has been made several times that there has been extraordinary success in obtaining a large number of recessive mutants of widely differing phenotype in quasidiploid cells. One hypothesis that has been offered to explain these unexpected results is that continuous lines during their evolution have become functionally hemizygous at the loci involved, either by chromosome deletions or chromosome loss, by rearrangements and loss of func-

tional transcription, or by some other process of gene inactivation[1,2,240]. Although there is some evidence for this hypothesis, little is known about the exact mechanisms[1,131,241]. In fact since it is probable that genes for the purine pathways or aminoacyl tRNA synthetases are scattered throughout the genome, it is extraordinary that such a large fraction would have become functionally hemizygous.

Recent work in our laboratory and in those of Farber and of Thompson have provided an extension of this hypothesis explaining the success in obtaining recessive isolates[65,242]. In our laboratory we have shown that emetine resistance in V79 cells arises in two steps, one involving a relatively high frequency event (10^{-2} in the absence of mutagenesis) in which the cells may become hemizygous, and secondly, a low frequency event (10^{-5} after mutagenesis) in which the remaining functional allele is mutated to emetine resistance. The molecular mechanism of the primary event is unknown but preliminary evidence suggests that it is a reversible process and thus is unlikely to involve gross chromosomal translocations or deletions. Further evidence for recessive mutations arising from a two step process has been obtained by Thompson and his group for an aminoacyl-tRNA synthetase[65], and by Farber for adenosine kinase[242]. It is conceivable therefore that most of the recessive mutants originate in this way.

(2) It is clear that there are now several systems which are good candidates for gene transfer experiments. These include the conditional lethals, where the wild type acts as the dominant allele, and many of the dominant isolates which we have described in the Tables. In the latter category, successful transfer has been described for the following dominant systems, dihydrofolate reductase[112,113], ribonucleotide reductase[112], Na^+/K^+ ATPase (ouabain resistance)[242], and asparagine synthetase[157].

(3) A large fraction of the mutants have been obtained in Chinese hamster cells, particularly the CHO line. It is not clear whether this line is particularly "susceptible", perhaps because of extensive functional hemizygosity, or whether insufficient efforts have been expended in other lines. However the absence of appropriate mutation systems in other lines may limit gene transfer work since CHO does not seem to be the best line to use for those purposes.

(4) Many of the mutant systems that have been developed mimic human diseases. These may significantly expedite our understanding of the molecular basis of the disease.

(5) Since most of the systems we have described are susceptible to mutagenesis, and many are amenable to two-way selection,

it is clear that there are now several systems which can be used
for studies in quantitative mutagenesis.

(6) Although there is good evidence that many of the mutants
represent structural gene mutations or gene amplification, the
evidence for this conclusion is not as clear in all systems. For
example, although the lesions have been identified in many of the
lectin resistant mutants, the frequencies in CHO cells after mutagenesis for some of the isolates is very high, the lesion in at
least one case is all or none (Pha^{RI}), and no reversion has been
demonstrated. Similarly in the toyocamycin resistance adenosine
kinase minus mutants, the frequencies are very high in CHO cells
(10^{-3} without mutagenesis), the lesion is all or none, and no
reversion has been demonstrated. These systems may therefore represent other mechanisms leading to a mutant phenotype.

Thus although we have come a long way in this field over the
last 10 - 15 years, much is unknown about the mechanisms of mutation in somatic cells. Hopefully with the advent of new technologies, one may be able to examine these questions.

ACKNOWLEDGMENTS

We wish to acknowledge financial support from the Medical Research Council and National Cancer Institute of Canada.

REFERENCES

1. L. Siminovitch, On the origin of mutants of somatic cells, in
   "Eucaryotic Gene Regulation", ICN-UCLA Symposia on Molecular
   and Cellular Biology, R. Axel, T. Maniatis, and C.F. Fox,
   eds., Academic Press, New York, pp.433-443 (1979).

2. L. Siminovitch, The Nature of Gene Variation and Gene Transfer
   in Somatic Cells, in "Genes, Chromosomes, and Neoplasia", F.E.
   Arrighi, P.N. Rao, and E. Stubblefield, eds., Raven Press, New
   York (1981).

3. J. Hochstadt, H.L. Ozer and C. Shopsis, Genetic Alteration in
   Animal Cells in Culture, Current Topics, Microbiology and
   Immunology, W. Arber and H. Koprowsky, eds., Springer Verlag,
   Berlin, New York (in press).

4. P. Coffino, R. Baumal, R. Laskov and M.D. Scharff, Cloning of
   Mouse Myeloma Cells and Detection of Rare Variants, J. Cell
   Physiol. 79:429-440 (1971).

5. T.T. Puck and F.T. Kao, Genetics of somatic mammalian cells. V. Treatment with 5-bromodeoxyuridine and visible light for isolation of nutritionally deficient mutants. Proc. Natl. Acad. Sci. USA 58:1227-1234 (1967).

6. F.T. Kao and T.T. Puck, Genetics of somatic mammalian cells. VII. Induction and isolation of nutritional mutants in Chinese hamster cells, Proc. Natl. Acad. Sci. USA 60:1275-1281 (1968).

7. L.H. Thompson, R. Mankovitz, R.M. Baker, J.E. Till, L. Siminovitch and G.F. Whitmore, Isolation of temperature-sensitive mutants of L-cells, Proc. Natl. Acad. Sci. USA 66:377-384 (1970).

8. F.T. Kao and T.T. Puck, Genetics of somatic mammalian cells. IX. Quantitation of mutagenesis by physical and chemical agents, J. Cell. Physiol. 74:245-258 (1969).

9. F.T. Kao, L. Chasin and T.T. Puck, Genetics of somatic mammalian cells. X. Complementation analysis of glycine-requiring mutants, Proc. Natl. Acad. Sci. USA 64:1284-1291 (1969).

10. F.T. Kao and T.T. Puck, Genetics of Chinese hamster cell mutants with respect to the requirement for proline, Genetics 55:513-529 (1967).

11. D. Patterson, F.T. Kao and T.T. Puck, Genetics of somatic mammalian cells: Biochemical genetics of Chinese hamster cell mutants with deviant purine metabolism, Proc. Natl. Acad. Sci. USA 71:2057-2061 (1975).

12. D. Patterson, Biochemical genetics of Chinese hamster cell mutants with deviant purine metabolism: Biochemical analysis of eight mutants, Somat. Cell Genet. 1:91-110 (1975).

13. D. Patterson, Biochemical genetics of Chinese hamster cell mutants with deviant purine metabolism. III. Isolation and characterization of a mutant unable to convert IMP to AMP, Somat. Cell Genet. 2:41-53 (1976).

14. D. Patterson, Biochemical genetics of Chinese hamster cell mutants with deviant purine metabolism. IV. Isolation of a mutant which accumulates adenylosuccinic acid and succinyl-aminoimidazole carboxamide ribotide, Somat. Cell Genet. 2:189-203 (1976).

15. D.C. Oates and D. Patterson, Biochemical genetics of Chinese
    hamster cell mutants with deviant purine metabolism: Charac-
    terization of Chinese hamster cell mutants defective in phos-
    phoribosylpyrophosphate amidotransferase and phosphoribosyl-
    glycinamide synthetase and an examination of alternatives to
    the first step of purine biosynthesis, Somat. Cell Genet.
    3:561-577 (1977).

16. E.H.Y. Chu, N.C. Sun and C.C. Chang, Induction of auxotrophic
    mutations by treatment of Chinese hamster cells with
    5-bromodeoxyuridine and black light, Proc. Natl. Acad. Sci.
    USA 69:3459-3463 (1972).

17. R.K. Feldman and M.W. Taylor, Purine mutants of mammalian cell
    lines. II. Identification of a phosphoribosylpyrophosphate
    amidotransferase-deficient mutant of Chinese hamster lung
    cells, Biochem. Genet. 13:227-234 (1975).

18. R.K. Feldman and M.W. Taylor, Purine mutants of mammalian cell
    lines. I. Accumulation of formylglycinamide ribotide by pur-
    ine mutants of Chinese hamster ovary cell, Biochem. Genet.
    12:393-405 (1975).

19. E.W. Holmes, G.L. King, A. Layva and S.C. Singer, A purine
    auxotroph deficient in phosphoribosylpyrophosphate amidotrans-
    ferase activities with normal activity of ribose-5-phosphate
    aminotransferase, Proc. Natl. Acad. Sci. USA 73:2458-2461
    (1976).

20. R.T. Taylor and M.L. Hanna, Folate-dependent enzymes in cul-
    tured Chinese hamster cells: Folylpolyglutamate synthetase
    and its absence in mutants auxotrophic for glycine + adenosine
    + thymide, Arch. Biochem. Biophys. 181:331-344 (1977).

21. M. McBurney and G.F. Whitmore, Isolation and biochemical char-
    acterization of folate deficient mutants of Chinese hamster
    cells, Cell 2:173-182 (1974).

22. M. McBurney and G.F. Whitmore, Characterization of a Chinese
    hamster cell with a temperature-sensitive mutation in folate
    metabolism, Cell 2:183 (1974).

23. G. Urlaub and L.A. Chasin, Isolation of Chinese hamster cell
    mutants deficient in dihydrofolate reductase activity, Proc.
    Natl. Acad. Sci. USA 77:4216-4220 (1980).

24. M. Meuth, M. Trudel and L. Siminovitch, Selection of Chinese
    hamster cells auxotrophic for thymidine by 1-B-D-arbinofurano-
    syl cytosine, Somat. Cell Genet. 5:303-318 (1979).

25. D. Ayusawa, H. Koyama, K. Iwata and T. Seno, Single-step se-
    lection of mouse FM3A cell mutants defective in thymidylate
    synthetase, Somat. Cell Genet. 6:261-270 (1980).

26. D. Patterson and D.V. Carnright, Biochemical genetic analysis
    of pyrimidine biosynthesis in mammalian cells. I. Isolation
    of a mutant defective in the early steps of de novo pyrimidine
    synthesis, Somat. Cell Genet. 3:483-495 (1977).

27. J.N. Davidson, D.V. Carnright and D. Patterson, Biochemical
    genetic analysis of pyrimidine biosynthesis in mammalian
    cells. III. Association of carbamyl phosphate synthetase,
    aspartate transcarbamylase, and dihydroorotase in mutants of
    cultured Chinese hamster cells, Somat. Cell Genet. 5:176-191
    (1979).

28. T. Kusano, M. Kato and I. Yamane, Isolation of uridine requir-
    ing variants in Chinese hamster cells. Cell Struct. and
    Funct. 1:393-396 (1976).

29. D. Patterson, Isolation and characterization of 5-fluoro-
    uracil-resistant mutants of Chinese hamster ovary cells defic-
    ient in the activities of orotate phosphoribosyltransferase
    and orotine 5'-monophosphate decarboxylase, Somat. Cell Genet.
    6:101-114 (1980).

30. B.B. Levinson, B. Ullman and D.W. Martin, Jr., Pyrimidine
    pathway variants of cultured mouse lymphoma cells with altered
    levels of both orotate phosphoribosyltransferase and orotidy-
    late decarboxylase, J. Biol. Chem. 254:4396-4404 (1979).

31. R.S. Krooth, W-L. Hsiao and B.W. Potvin, Resistance to
    5-fluoroorotic acid and pyrimidine auxotrophy: A new bidirec-
    tional selective system for mammalian cells, Somat. Cell
    Genet. 5:551-569 (1979).

32. Y. Saito, S.M. Chou and D.F. Silbert, Animal cell mutants
    defective in sterol metabolism: A specific selection proce-
    dure and partial characterization of defects, Proc. Natl.
    Acad. Sci. USA 74:3730-3734 (1977).

33. T-Y. Chang and P.R. Vagelos, Isolation and characterization of
    an unsaturated fatty acid-requiring mutant of cultured mammal-
    ian cells, Proc. Natl. Acad. Sci. USA 73:24-28 (1976).

34. T-Y. Chang, C. Telakowski, W. Vanden Heuvel, A.W. Alberts and
    P.R. Vagelos, Isolation and partial characterization of a
    cholesterol-requiring mutant of Chinese hamster ovary cells,
    Proc. Natl. Acad. Sci. USA 74:832-836 (1977).

35. J.S. Limanek, J. Chn and T-Y. Chang, Mammalian cell mutant requiring cholesterol and unsaturated fatty acid for growth, Proc. Natl. Acad. Sci. USA 75:5452-5456 (1978).

36. K. Hidaka, S-I. Akiyama and M. Kuwano, Growth of amphotericin B-resistant hamster cell line requires exogenous cholesterol, Exper. Cell Res. 128:215-221 (1980).

37. C. Jones and T.T. Puck, Genetics of somatic mammalian cells. XVII. Induction and isolation of Chinese hamster cell mutants requiring serine, J. Cell Physiol. 81:299-304 (1973).

38. O. Hankinson, Mutants of the Chinese hamter ovary cell line requiring alanine and glutamate, Somat. Cell Genet. 2:497-507 (1976).

39. M.M.Y. Waye and C.P. Stanners, Isolation and characterization of CHO cell mutants with altered asparagine synthetase, Somat. Cell Genet. 5:625-639 (1979).

40. G. Ditta, K. Soderberg, F. Landy and I.E. Scheffler, The Selection of Chinese hamster cells deficient in oxidative energy metabolism, Somat. Cell Genet. 2:331-344 (1976).

41. L. DeFrancesco, D. Werntz and I.E. Scheffler, Conditionally lethal mutations in Chinese hamster cells. Characterization of a cell line with a possible defect in the Krebs cycle. J. Cell. Physiol. 85:293-306 (1975).

42. S.L. Naylor, L.L. Busby and R.J. klebe, Biochemical selection systems for mammalian cells: The essential amino acids, Somat. Cell Genet. 2:93-111 (1976).

43. S.L. Naylor, J.K. Townsend and R.J. Klebe, Characterization of naturally occurring auxotrophic mammalian cells, Somat. Cell Genet. 5:271-277 (1979).

44. P.F. Coleman, D.P. Suttle and G.R. Stark, Purification from hamster cells of the multifunctional protein that initiates de novo synthesis of pyrimidine nucleotides, J. Biol. Chem. 252:6379-6385 (1977).

45. T. Kempe, E. Swyrd, M. Bruist and G. Stark, Stable mutants of mammalian cells that overproduce the first three enzymes of pyrimidine nucleotide biosynthesis, Cell 9:541-550 (1976).

46. R.A. Padgett, G.M. Wahl, P.F. Coleman and G.R. Stark, N-(phos-phonacetyl)-L-aspartate-resistant hamster cells over-accumu-late a single mRNA coding for the multifunctional protein that catalyzes the first step of UMP synthesis, J. Biol. Chem. 254:974-980 (1979).

47. L.A. Chasin, A. Feldman, M. Konstam and G. Urlaub, Reversion of a Chinese hamster cell auxotrophic mutant, Proc. Natl. Acad. Sci. USA 71:718 (1974).

48. D.C. Oates, D. Vannais and D. Patterson, A. mutant of CHO-K1 cells deficient in two-nonsequential steps of de novo purine biosynthesis, Cell 20:797-805 (1980).

49. D. Patterson, S. Graw and C. Jones, Demonstration, by somatic cell genetics, of coordinate regulation of genes for two en-zymes of purine synthesis assigned to human chromosome 21, Proc. Natl. Acad. Sci. USA 78:405-409 (1981).

50. P.M. Naha, Temperature sensitive conditional mutants of monkey kidney cells, Nature 223:1380-1381 (1969).

51. B.J. Smith and N.M. Wigglesworth, A cell line which is temper-ature-sensitive for cytokinesis, J. Cell Physiol. 80:253-260 (1972).

52. H.K. Meiss and C. Basilico, Temperature-sensitive mutants of BHK 21 cells, Nature New Biol. 239:66-68 (1972).

53. C. Basilico, Temperature-sensitive mutations in animal cells, Adv. Can. Res. 24:223-266 (1977).

54. L.H. Thompson, J.L. Harkins and C.P. Stanners, A mammalian cell mutant with a temperature-sensitive leucyl-transfer RNA synthetase, Proc. Natl. Acad. Sci. USA 70:3094-3098 (1973).

55. L.H. Thompson, C.P. Stanners and L. Siminovitch, Selection by [^{3} H] amino acids of CHO-cell mutants with altered leucyl and asparagyl-transfer RNA synthetases, Somat. Cell Genet. 1:187-208 (1975).

56. R.A. Farber and M.P. Deutscher, Physiological and biochemical properties of a temperature-sensitive leucyl-tRNA synthetase mutant (tsHI) and revertant from Chinese hamster cells, Somat. Cell Genet. 2:509-520 (1976).

57. L. Haars, A. Hampel and L.H. Thompson, Altered leucyl-transfer RNA synthetase from a mammalian cell culture mutant, Biochim. Biophys. Acta 454:493-503 (1976).

58. J.J. Wasmuth and C.T. Caskey, Selection of temperature-sensitive CHL asparagyl t-RNA synthetase mutants using the toxic lysine analog, S-2-aminoethyl-L-cysteine, Cell 9:655-662 (1976).

59. L.H. Thompson, D.J. Lofgren and G.M. Adair, CHO cell mutants for arginyl-, asparagyl-, glutaminyl-, histidyl-, and methionyl-transfer RNA synthetases: Identification and initial characteization, Cell 11:157-168 (1977).

60. L.H. Thompson, D.J. Lofgren and G.M. Adair, Evidence for structural gene alterations affecting aminoacyl-tRNA synthetases in CHO cell mutants and revertants, Somat. Cell Genet. 4:423-435 (1978).

61. G.M. Adiar, L.H. Thompson and P.A. Lindl, Six complementation classes of conditionally lethal protein synthesis mutants of CHO cells selected by ^{3}H-amino acids, Somat. Cell Genet. 4:27-44 (1978).

62. I.L. Andrulis, G.S. Chiang, S.M. Arfin, T.A. Miner and G.W. Hatfield, Biochemical characterization of a mutant asparaginyl-tRNA synthetase from Chinese hamster ovary cells, J. Biol. Chem. 253:58-62 (1978).

63. C.R. Ashman, Mutations in the structural genes of CHO cell histidyl-, valyl-, and leucyl-tRNA synthetases, Somat. Cell Genet. 4:299-312 (1978).

64. C.P. Stanners, T.M. Wightman and J.L. Harkins, Effect of extreme amino acid starvation on the protein synthetic machinery of CHO cells, J. Cell Physiol. 95:125-138 (1978).

65. G.M. Adair, L.H. Thompson and S. Fond, [^{3}H] Amino acid selection of aminoacyl-tRNA synthetase mutants of CHO cells: Evidence of homovs. hemizygosity at specific loci, Somat. Cell Genet. 5:329-344 (1979).

66. D. Toniolo, H.K. Meiss and C. Basilico, A temperature-sensitive mutation affecting 28S ribosomal RNA production in mammalian cells, Proc. Natl. Acad. Sci. USA 70:1273-1277 (1973).

67. D. Toniolo and C. Basilico, Processing of ribosomal RNA in a temperature sensitive mutant of BHK cells, Biochim. Biophys. Acta 425:409-418 (1976).

68. L.H. Thompson, R. Mankovitz, R.M. Baker, J.A. Wright, J.E.
    Till, L. Siminovitch and G.F. Whitmore, Selective and non-
    selective isolation of temperature sensitive mutants of mouse
    L-cells and their characterization, J. Cell Physiol. 78:431-
    439 (1971).

69. M.L. Slater and H.L. Ozer, Temperature-sensitive mutants of
    Balb/3T3 cells: Description of a mutant affected in cellular
    and polyoma virus DNA synthesis, Cell 7:289-295 (1976).

70. R. Sheinin, Preliminary characterization of the temperature-
    sensitive defect in DNA replication in a mutant mouse L cell,
    Cell 7:49-57 (1976).

71. K.K. Jha and H. Ozer, Genetic studies with a mutant mouse
    cell, ts-2 Balb/3T3 with a temperature-sensitive defect in DNA
    synthesis, Genetics (suppl.) 86:532-534 (1977).

72. D.J. Roufa, S.M. McGill and J.W. Mollenkamp, Ts mutant iso-
    lated from CHO cells-inhibition of DNA-replication at non-
    permissive temperature, Somat. Cell Genet. 5:97-115 (1979).

73. I.E. Scheffler and G. Buttin, Conditionally lethal mutations
    in Chinese hamster cells. I. Isolation of a temperature-
    sensitive line and its investigation by cell cycle studies, J.
    Cell Physiol. 81:199-216 (1973).

74. P.M. Naha, A.L. Meyer and K. Hewitt, Mapping of the G_1 phase
    of mammalian cell cycle, Nature 258:49 (1975).

75. R.J. Wang, A novel temperature-sensitive mammalian cell line
    exhibiting defective prophase progression, Cell 8:257-261
    (1976).

76. C. Basilico, Selective production of cell cycle specific ts
    mutants, J. Cell Physiol. 95:367-376 (1978).

77. N.B. Liskay and D.M. Prescott, Genetic analysis of the G_1
    period: Isolation of mutants (or variants) with a G_1 period
    from a Chinese hamster cell line lacking G_1, Proc. Natl.
    Acad. Sci. USA 75:2873-2877 (1978).

78. H.K. Meiss, A. Talavera and T. Nishimoto, A recurring tempera-
    ture-sensitive mutant class of BHK-21 cells, Somat. Cell
    Genet. 4:125-130 (1978).

79. T. Nishimoto and C. Basilico, Analysis of a method for selec-
    ting temperature sensitive mutants of BHK-cells, Somat. Cell
    Genet. 4:323-340 (1978).

80. J. Melero, Isolation and cell cycle analysis of temperature-sensitive mutants from Chinese hamster cells, J. Cell Physiol. 98:17-30 (1979).

81. H.E. Schwartz, G.C. Moser, S. Holmes and H.K. Meiss, Assignment of temperature-sensitive mutations of BHK cells to the X chromosome, Somat. Cell Genet. 5:217-224 (1979).

82. P.M. Naha and R. Sorrentino, Biochemical controls of the G_1 phase of a mammalian cell cycle. I. Analysis of chromatin proteins in temperature-sensitive variants, Cell Biol. International Reports 4:365-378 (1980).

83. T. Nishimoto, T. Takahashi and C. Basilico, A temperature-sensitive mutation affecting S-phase progression can lead to accumulation of cells with a G-2 DNA content, Somat. Cell Genet. 6:465-476 (1980).

84. R. Fenwick, Jr. and C. Caskey, Mutant Chinese hamster cells with a thermosensitive hypoxanthine-guanine phosphoribosyl-transferase, Cell 5:115-122 (1975).

85. R.G. Fenwick, Jr., T.H. Sawyer, G.D. Kruh, K.H. Astrin and C.T. Caskey, Forward and reverse mutations affecting the kinetics and apparent molecular weight of mammalian HGPRT, Cell 12:383-391 (1977).

86. C.J. Ingles, Temperature sensitive RNA polymerase II mutations in Chinese hamster ovary cells, Proc. Natl. Acad. Sci. USA 75:405-409 (1978).

87. M.M. Nakano, T. Sekiquicki and M. Hamada, A mammalian cell mutant with temperature-sensitive thymidine kinase, Somat. Cell Genet. 4:169-178 (1978).

88. K. Miyashita and T. Kakunaga, Isolation of heat and cold-sensitive mutants of Chinese hamster lung cells affected in their ability to express the transformed state, Cell 5:131-138 (1975).

89. R.A. Farber and P. Unrau, Isolation of cold-sensitive Chinese hamster cells, Molec. Genet. 138:233-242 (1975).

90. M.S.J. Crane and D.B. Thomas, Cell-cycle, cell-shape mutant with features of the G_0 state, Nature 261:205-208 (1976).

91. B.M. Ohlsson-Wilhelm, J.F. Leary, M. Pacilio and T. Martin, Rapid, quantitative analysis of cell cycle stages of cold-sensitive derivatives of the Chinese hamster cell line CHO-K1, Somat. Cell Genet. 6:349-359 (1980).

92. V. Ling, A membrane-altered mutant cold-sensitive for growth, J. Cell Physiol. 91:209-224 (1977).

93. T-S. Chan, C. Long and H. Green, A human-mouse somatic hybrid line selected for human deoxycytidine deaminase, Somat. Cell Genet. 1:81-90 (1975).

94. B. Ullman, L.H. Gudas, S.M. Clift and D.W. Martin, Jr., Isolation and characterization of purine-nucleoside phosphorylase-deficient T-lymphoma cells and secondary mutants with altered ribonucleotide reductase: Genetic model for immunodeficiency disease, Proc. Natl. Acad. Sci. USA 76:1074-1078 (1979).

95. P. Hoffee, A method for the isolation of purine nucleoside phosphorylase-deficient variants of mammalian cell lines, Somat. Cell Genet. 5:319-328 (1979).

96. E.R. Giblett, A.J. Ammann, D.W. Wara, R. Sandman and L.K. Diamond, Nucleoside-phosphorylase deficiency in a child with severely defective T-cell immunity and normal B-cell immunity, The Lancet 1:1010-1013 (1975).

97. R.L. Momparler, M.Y. Chu and G.A. Fischer, Studies on a new mechanism of resistance of L5178Y murine leukemia cells to cytosine arabinoside, Biochim. Biophys. Acta 161:481-493 (1968).

98. B.R. de Saint-Vincent and G. Buttin, Studies on 1-β-D-Arbinofuranosyl-cytosine-resistant mutants of Chinese hamster fibroblasts, Eur. J. Biochem. 37:481-488 (1973).

99. M. Meuth and H. Green, Alterations leading to increased ribonucleotide reductase in cells selected for resistance to deoxynucleosides, Cell 3:367-374 (1974).

100. M. Meuth, E. Aufreiter and P. Reichard, Deoxyribonucleotide pools in mouse-fibroblast cell lines with altered ribonucleotide reductase, Eur. J. Biochem. 71:39-43 (1976).

101. M. Meuth, N. L'Heureux-Huard and M. Trudel, Characterization of a mutator gene in Chinese hamster ovary cells, Proc. Natl. Acad. Sci. USA 76:6506-6509 (1979).

102. V.L. Chan and P. Juranka, Isolation and preliminary characterization of 9-β-D-arabinofuranosyladenine-resistant mutants of baby hamster cells, Somat. Cell Genet. 7:147-160 (1981).

103. W.F. Flintoff, S.V. Davidson and L. Siminovitch, Isolation and partial characterization of three methotrexate-resistant phenotypes from Chinese hamster ovary cells, Somat. Cell Genet. 2:245-261 (1976).

104. W.F. Flintoff, S.M. Spindler and L. Siminovitch, Genetic characterization of methotrexate-resistant Chinese hamster ovary cells, In Vitro 12:749-757 (1976).

105. F.W. Alt, R.E. Kellems and R.T. Schimke, Synthesis and degradation of folate reductase in sensitive and methotrexate-resistant lines of S-180 cells, J. Biol. Chem. 251:3063-3074 (1976).

106. R.S. Gupta, W.F. Flintoff and L. Siminovitch, Purification and properties of dihydrofolate reductase from methotrexate-sensitive and resistant Chinese hamster ovary cells, Canad. J. Biochem. 55:445-452 (1977).

107. R.T. Schimke, R.J. Kaufman, F.W. Alt and R.F. Kellems, Gene amplification and drug resistance in cultured murine cells, Science 202:1051-1055 (1978).

108. J.H. Nunberg, R.J. Kaufman, R.T. Schimke, G. Urlaub and L.A. Chasin, Amplified dihydrofolate reductase genes are localized to a homogeneously staining region of a single chromosome in a methotrexate-resistant Chinese hamster ovary cell line, Proc. Natl. Acad. Sci. USA 75:5553-5556 (1978).

109. R.J. Kaufman, P.C. Brown and R.T. Schimke, Amplified dihydrofolate reductase genes in unstably methotrexate-resistant cells are associated with double minute chromosomes, Proc. Natl. Acad. Sci. USA 76:5669-5673 (1979).

110. R.E. Kellems, V.B. Morhenn, E.A. Pfendt, F.W. Alt and R.T. Schimke, Polyoma virus and cyclic AMP-mediated control of dihydrofolate reductase mRNA abundance in methotrexate-resistant mouse fibroblasts, J. Biol. Chem. 254:309-318 (1979).

111. B.J. Dolnick, R.J. Berenson, J.R. Bertino, R.J. Kaufman, J.H. Nunberg and R.T. Schimke, Correlation of dihydrofolate reductase elevation with gene amplification in a homogeneously staining chromosomal region in L5178Y cells, J. Cell Biol. 83:394-402 (1979).

112. W.H. Lewis, P.R. Srinivasan, N. Stokoe and L. Siminovitch, Parameters governing the transfer of the genes for thymidine kinase and dihydrofolate reductase into mouse cells using metaphase chromosomes of DNA, Somat. Cell Genet. 6:333-348 (1980).

113. M. Wigler, M. Perucho, D. Kurtz, S. Dana, A. Pellicer, R. Axel and S. Silverstein, Transformation of mammalian cells with an amplifiable dominant-acting gene, Proc. Natl. Acad. Sci. USA 77:3567-3570 (1980).

114. J.P. Thirion, D. Banville and H. Noel, Galactokinase mutants of Chinese hamster somatic cells resistant to 2-deoxyglucose, Genetics 83:137-147 (1976).

115. J.A. Wright and W.H. Lewis, Evidence for a common site of action for the antitumor agents hydroxyurea and guanazole, J. Cell Physiol. 83:437-440 (1974).

116. W.H. Lewis and J.A. Wright, Altered ribonucleotide reductase activity in mammalian tissue culture cells resistant to hydroxyurea, Biochem. Biophys. Res. Commun. 60:926-933.

117. W.H. Lewis and J.A. Wright, Genetic characterization of hydroxyurea-resistance in Chinese hamster ovary cells, J. Cell Physiol. 97:73-86 (1978).

118. W.H. Lewis and J.A. Wright, Isolation of hydroxyurea-resistant cells with altered levels of ribonucleotide reductase, Somat. Cell Genet. 5:83-96 (1979).

119. R.L. Davidson and E.R. Kaufman, Resistance to bromodeoxy-uridine mutagenesis and toxicity in mammalian cells selected for resistance to hydroxyurea, Somat. Cell Genet. 5:873-885 (1979).

120. C-C. Chang, J.A. Boezi, S.T. Warren, C.L.K. Sabourin, P.K. Liu, L. Glatzer and J.E. Trosko, Isolation and characterization of a UV-sensitive hypermutable aphidicolin-resistant Chinese hamster cell line, Somat. Cell Genet. 7:235-253 (1981).

121. C.L.K. Sabourin, P.F. Bates, L. Glatzer, C-C. Chang, J.E. Trosko and J.A. Boezi, Selection of aphidicolin-resistant CHO cells with altered levels of ribonucleotide reductase, Somat. Cell Genet. 7:255-268 (1981).

122. M.W. McBurney and G.F. Whitmore, Mutants of Chinese hamster cells resistant to adenosine, J. Cell. Physiol. 85:87-100 (1975).

123. R.S. Gupta and L. Siminovitch, Genetic and biochemical studies with the adenosine analogs toyocamycin and tubercidin: Mutation at the adenosine kinase locus in Chinese hamster cells, Somat. Cell Genet. 4:715-736 (1978).

124. T-S. Chan, R.P. Creagan and M.P. Reardon, Adenosine kinase as a new selective marker in somatic cell genetics: Isolation of adenosine kinase-deficient mouse cell lines and human-mouse hybrid cell lines containing adenosine kinase, Somat. Cell Genet. 4:1-12 (1978).

125. M.S. Rabin and M.M. Gottesman, High frequency of mutation to tubercidin resistance in CHO cells, Somat. Cell Genet. 5:571-583 (1979).

126. W. Szbalski and M.J. Smith, Genetics of human cell lines. I. 8-azaguanine resistance, a selective "single-step" marker, Proc. Soc. Exp. Biol. Med. 101:662-666 (1959).

127. J.W. Littlefield, The inosinic acid pyrophosphorylase activity of mouse fibroblasts partially resistant to 8-azaguanine, Proc. Natl. Acad. Sci. USA 50:568-576 (1963).

128. E.H.Y. Chus, P. Brimer, K.B. Jacobsen and E. Merriam, Mammalian cell genetics. I. Selection and characterization of mutations auxotrophic for L-glutamine or resistant to 8-azaguanine in Chinese hamster cells in vitro, Genetics 62:359-377 (1969).

129. F.D. Gillin, D.J. Roufa, A.L. Beaudet and C.T. Caskey, 8-Azaguanine resistance in mammalian cells. I. Hypoxanthine-guanine phosphoribosyltransferase, Genetics 72:239-252 (1972).

130. J.D. Sharp, N.E. Capecchi and M.R. Capecchi, Altered enzymes in drug-resistant variants of mammalian tissue culture cells, Proc. Natl. Acad. Sci. USA 70:3145-3149 (1973).

131. L.A. Chasin, Mutations affecting adenine phosphoribosyl transferase activity in Chinese hamster cells, Cell 2:43-54 (1974).

132. G.E. Jones and P.A. Sargent, Mutants of cultured Chinese hamster cells deficient in adenine phosphoribosyl transferase, Cell 2:37-41 (1974).

133. M.W. Taylor, J.H. Pipkorn, M.K. Tokito and R.O. Pozzatti, Jr., Purine mutants of mammalian cells. III. Control of purine biosynthesis in adenine phosphoribosyl transferase mutants of CHO cells, Somat. Cell Genet. 3:195-206 (1977).

134. S. Kit, D.R. Dubbs, L.H. Piekarski and T.C. Hsu, Deletion of thymidine kinase activity from L cells resistant to bromodeoxyuridine, Exp. Cell Res. 31:297-312 (1963).

135. J.W. Littlefield, Studies on thymidine kinase in cultured mouse fibroblasts, Biochim. Biophys. Acta 95:14-22 (1965).

136. D. Clive, W.G. Flamm, M.R. Machesko and N.J. Berhneim, A muta-
     tional assay system using the thymidine kinase locus in mouse
     lymphoma cells, Mutat. Res. 16:77-87 (1972).

137. E.R. Kaufman and R.L. Davidson, Novel phenotypes arising from
     selection of hamster melanoma cells for resistance to BUdR,
     Exp. Cell Res. 107:15-24 (1977).

138. E.R. Kaufman and R.L. Davidson, Effects of thymidine analogs
     on Syrian hamster melanoma cells: Phenotypes arising from
     selection to analog resistance, Somat. Cell Genet. 3:649-661
     (1977).

139. L. Medrano and H. Green, A uridine kinase-deficient mutant of
     3T3 and a selective method for cells containing the enzyme,
     Cell 1:23-26 (1974).

140. M. Greenbert, D.E. Shumm and T.E. Weeb, Uridine kinase activi-
     ties and pyrimidine nucleoside phosphorylation in fluoropyri-
     midine sensitive and resistant cell lines of the Novikoff
     Hepatoma, Biochem. J. 164:379-387 (1979).

141. G.M. Wahl, R.A. Padgett and G.R. Stark, Gene amplification
     causes overproduction of the first three enzymes of UMP syn-
     thesis in N-(Phosphonacetyl)-L-aspartate-resistant hamster
     cells, J. Biol. Chem. 254:8679-8689 (1979).

142. D.P. Suttle and G.R. Stark, Coordinate overproduction of oro-
     tate phosphoribosyltransferase and orotidine-5'-phosphate
     decarboxylase in hamster cells resistant to pyrazofurin and
     6-azauridine, J. Biol. Chem. 254:4602-4607 (1979).

143. M. Wigler, S. Silverstein, L-S. Lee, A. Pellicer, Y-C. Cheng
     and R. Axel, Transfer of purified herpes virus thymidine ki-
     nase gene to cultured mouse cells, Cell 11:223-232 (1977).

144. A.C. Minson, P. Wildy, A. Buchan and G. Darby, Introduction of
     the herpes simplex virus thymidine kinase gene into mouse
     cells using virus DNA or transformed cell DNA, Cell 13:581-587
     (1978).

145. M. Wigler, A. Pellicer, S. Silverstein, R. Axel, G. Urlaub and
     L. Chasin, DNA-mediated transfer of the adenine phosphori-
     bosyl-transferase locus into mammalian cells, Proc. Natl.
     Acad. Sci. USA 76:1373-1376 (1979).

146. M. Wigler, R. Sweet, G.K. Sim, B. Wold, A. Pellicer, E. Lacy,
     T. Maniatis, S. Silverstein and R. Axel, Transformation of
     mammalian cells with genes from procaryotes and eucaryotes,
     Cell 16:777-785 (1979).

147. L.H. Graf, Jr., G. Urlaub and L.A. Chasin, Transformation of the gene hypoxanthine phosphoribosyltransferase, Somat. Cell Genet. 5:1031-1044 (1979).

148. K. Willecke, M. Klomfass, R. Mierau and J. Dohmer, Intra-species transer via total cellular DNA of the gene for hypo-xanthine phosphoribosyltransferase into cultured mouse cells, Mol. Gen. Genetics 179:179-185 (1979).

149. S.C. Lester, S.K. LeVan, C. Steglich and R. DeMars, Expression of human genes for adenine phosphoribosyltransferase and hypo-xanthine-guanine phosphoribosyltransferase after genetic transformation of mouse cells with purified human DNA, Somat. Cell Genet. 6:241-259 (1980).

150. J.L. Peterson and O.W. McBride, Cotransfer of linked eukary-otic genes and efficient transfer of hypoxanthine phosphoribo-syltransferase by DNA-mediated gene transfer, Proc. Natl. Acad. Sci. USA 77:1583-1587 (1980).

151. R.M. Liskay and R.J. Evans, Inactive X chromosome DNA does not function in DNA-mediated cell transformation for the hypoxan-thine phosphoribosyltransferase gene, Proc. Natl. Acad. Sci. USA 77:4895-4898 (1980).

152. L.A. Chasin and G. Urlaub, Mutant alleles for hypoxanthine phosphoribosyl transferase: Codominant expression, complemen-tation and segregation in hybrid Chinese hamster cells, Somat. Cell Genet. 2:453-467 (1976).

153. S.A. Farrell and R.G. Worton, Chromosome loss is responsible for segregation at the HPRT locus in Chinese hamster cell hybrids, Somat. Cell Genet. 3:539-551 (1977).

154. V. Ling, Drug resistance and membrane alteration in mutants of mammalian cells, Can. J. Genetics and Cytology 17:503-515 (1975).

155. R.L. Juliano and V. Link, A surface glycoprotein modulating drug permeability in Chinese hamster ovary cell mutants, Biochim. Biophys. Acta 455:152-162 (1976).

156. R.L. Juliano, J. Graves and V. Link, Drug-resistant mutants of Chinese hamster ovary cells possess an altered cell surface carbohydrate component, J. Supramolec. Struct. 4:521-526 (1976).

157. I.L. Andrulis and L. Siminovitch, DNA-mediated gene transfer of β-aspartylhydroxamate resistance into Chinese hamster ovary cells, Proc. Natl. Acad. Sci. USA 78:5724-5728 (1981).

158. I.L. Andrulis and L. Siminovitch, Isolation and characteriza-
     tion of CHO cells resistant to the amino acid analogue β-
     aspartyl hydroxamate (in preparation).

159. J. Choi and I.E. Scheffler, A mutant of Chinese hamster ovary
     cells resistant to α-methyl and α-difluoromethylornithine
     Somat. Cell Genet. 7:219-233 (1981).

160. M. Debatisse, M. Berry and G. Buttin, The potentiation of
     adenine toxicity to Chinese hamster cells by coformycin:
     suppression in mutants with altered regulation of purine bio-
     synthesis or increased adenylate-deaminase activity, J. Cell
     Physiol. 106:1-11 (1981).

161. M. Sinensky, Isolation of a mammalian cell mutant resistant to
     25-hydroxy cholesterol, Biochem. and Biophys. Res. Commun.
     78:863-867 (1977).

162. W.J. Rigby, B.D. Burleigh, Jr. and B.S. Hartley, Gene duplica-
     tion in experimental enzyme evolution, Nature 251:200-204
     (1974).

163. J.A. Wright, Pleiotrophic changes in lines of Chinese hamster
     ovary cells resistant to concanavalin and phytohaemagglutinin,
     J. Cell Biol. 56:666-675 (1972).

164. C. Gottlieb, A.M. Skinner and S. Kornfeld, Isolation of a
     clone of Chinese hamster ovary cells deficient in plant
     lectin-binding sites, Proc. Natl. Acad. Sci. USA 72:1078-1082
     (1974).

165. C. Gottlieb, J. Baenziger and S. Kornfeld, Deficient uridine
     diphosphate-N-acetylglucosamine: Glycoprotein N-acetyl-gly-
     cosaminyl-transferase activity in a clone of Chinese hamster
     ovary cells with altered surface glycoproteins, J. Biol. Chem.
     250:3303-3309 (1975).

166. R.L. Juliano and P. Stanley, Altered cell surface glycopro-
     teins phytohemagglutinin-resistant mutants of Chinese hamster
     ovary cells, Biochim. Biophys. Acta 389:401-406 (1975).

167. P. Stanley, V. Caillibot and L. Siminovitch, Stable altera-
     tions at the cell membrane of Chinese hamster ovary cells
     resistant to the cytotoxicity of phytohemagglutinin, Somat.
     Cell Genet. 1:3-26 (1975).

168. P. Stanley, V. Caillibot and L. Siminovitch, Selection and
     characterization of eight phenotypically-distinct lines of
     lectin-resistant Chinese hamster ovary cells, Cell 6:121-128
     (1975).

169. P. Stanely, S. Narasimhan, L. Siminovitch and H. Schachter, Chinese hamster ovary cells selected for resistance to the cytotoxicity of phytohemagglutinin are deficient in a UDP-N-acetyl-glucosamine-glycoprotein-N-acetyl-glucosaminyl-transferase activity, Proc. Natl. Acad. Sci. USA 72:3323-3327 (1975).

170. P. Stanley and L. Siminovitch, Complementation between mutants of CHO cells resistant to a variety of plant lectins, Somat. Cell Genet. 3:391-405 (1977).

171. S.S. Krag, M. Cifone, P.W. Robbins and R.M. Baker, Reduced synthesis of [^{14}C] mannosyl oligosaccharide-lipid by membranes prepared from concanavalin A-resistant Chinese hamster ovary cells, J. Biol. Chem. 252:3561-3564 (1977).

172. E.G. Briles, E. Li and S. Kornfeld, Isolation of wheat-germ agglutinin-resistant clones of CHO cells deficient in membrane sialic acid and galactose, J. Biol. Chem. 252:1107-1116 (1977).

173. P. Stanley, Surface carbohydrate alterations of mutant mammalian cells selected for resistance to plant lectins, in "Biochemistry of Glycoproteins and Proteoglycans," W.J. Lennarz, ed., Plenum Publishing Corporation, pp. 161-189 (1980).

174. T. Sudo and K. Onodera, Isolation and characterization of Tunicamycin resistant mutants from Chinese hamster ovary cells, J. Cell Physiol. 101:149-156 (1979).

175. M. Taub and E. Englesberg, Isolation and characterization of 5-fluorotryptophan-resistant mutants with altered L-tryptophan transport, Somat. Cell Genet. 2:441-452 (1976).

176. M.Taub and E. Englesberg, 5-Fluorotryptophan resistant mutants affecting the A and L transport system in the mouse L cell line A9, J. Cell Physio. 97:477-486 (1978).

177. E. Gnglesberg, R. Bass and W. Heiser, Inhibition of the growth of mammalian cells in culture by amino acids and the isolation and characterization of L-phenylalanine-resistant mutants modifying L-phenylalanine transport, Somat. Cell Genet. 2:411-428 (1976).

178. M.C. Finkelstein, C.W. Slayman and E.A. Adelberg, Tritium suicide selection of mammalian cell mutants defective in the transport of neutral amino acids, Proc. Natl. Acad. Sci. USA 74:4549-4551 (1977).

179. C.E. Campbell, R.A. Gravel and R.G. Worton, Isolation and characterization of Chinese hamster cell mutants resistant to the cytotoxic effects of chromate, Somat. Cell Genet. 7:535-546 (1981).

180. J. Mandel and W.I. Flintoff, Isolation of mutant mammalian cells altered in polyamine transport, J. Cell Physiol. 97:335-344 (1978).

181. V. Ling and L.H. Thompson, Reduced permeability in CHO cells as a mechanism of resistance to colchicine, J. Cell Physiol. 83:103-116 (1974).

182. N.T. Bech-Hansen, J.E. Till and V. Ling, Pleiotropic phenotype of colchicine-resistant CHO cells: Cross-resistance and collateral sensitivity, Jr. Cell Physiol. 88:23-39 (1976).

183. V. Ling and R.M. Baker, Dominance of colchicine resistance in hybrid CHO cells, Somat. Cell Genet. 6:193-200 (1978).

184. R.M. Baker, D.M. Brunette, R. Mankovitz, L.H. Thompson, G.F. Whitmore, L. Siminovitch and J.E. Till, Ouabain resistant mutants of mouse and hamster cells in culture, Cell 1:9-21 (1974).

185. C.H. Sibley and G.M. Tomkins, Isolation of lymphoma cell variant resistant to killing by glucocorticoids, Cell 2:213-220 (1974).

186. C.H. Sibley and G.M. Tomkins, Mechanisms of steroid resistance, Cell 2:221-227 (1974).

187. S. Bourgeois and R.F. Newby, Genetic analysis of glucocorticoid action on lymphoid cell lines, in "Hormones and Cancer," S. Iacobelli et al., eds., Raven Press, New York, pp. 67-77 (1980).

188. T.J. Moehring and J.M. Moehring, Selection and characterization of cells resistant to diphtheria toxin and pseudomonas exotoxin A: Presumptive translational mutants, Cell 11:447-454 (1977).

189. D. Pious, P. Hawley and G. Forrest, Isolation and characterization of HL-A variants in cultured human lymphoid cells, Proc. Natl. Acad. Sci. USA 70:1397-1400 (1973).

190. J.W. Dennis and R.S. Kerbel, Characterization of a deficiency in fucose metabolism in lectin-resistant variants of a murine tumor showing altered tumorigenic and metastatic capacities in vivo, Cancer Research 41:98-104 (1981).

191. J.W. Dennis, T.P. Donaghue and R.S. Kerbel, Membrane-associated alterations detected in poorly tumorigenic lectin-resistant variant sublines of a highly malignant and metastatic murine tumor, J. Natl. Cancer Inst. 66:129-139 (1981).

192. J. Dennis, T. Donaghue, M. Florian and R.S. Kerbel, Reversion of stable in vitro genetic markers detected in tumor cells from spontaneous metastases, Nature 292:242-245 (1981).

193. R.S. Gupta and L. Siminovitch, Isolation and preliminary characterization of mutants of CHO cells resistant to the protein synthesis inhibitor emetine, Cell 9:213-219 (1976).

194. R.S. Gupta and L. Siminovitch, The molecular basis of emetine resistance in Chinese hamster ovary cells: Alteration in the 40S subunit, Cell 10:61-66 (1977).

195. R.S. Gupta and L. Siminovitch, Mutants of CHO cells resistant to the protein synthesis inhibitors cryptopleurine and tylocrebrine: Genetic and biochemical evidence for common site of action of emetine, cryptopleurine, tylocrebrine and tubulosine, Biochem. 16:3209-3214 (1977).

196. R.S. Gupta and L. Siminovitch, Mutants of CHO cell resistant to the protein synthesis inhibitor emetine: Genetic and biochemical characterization of second step mutants, Somat. Cell Genet. 4:77-94 (1978).

197. R.S. Gupta and L. Siminovitch, An in vitro analysis of the dominance of emetine sensitivity in CHO cell hybrids, J. Biol. Chem. 253:3978-3982 (1978).

198. J.J. Wasmuth, J.M. Hill and L.S. Vock, Biochemical and genetic evidence for a new class of emetine-resistant Chinese hamster cells with alterations in the protein biosynthetic machinery, Somat. Cell Genet. 6:495-516 (1980).

199. R.S. Gupta and L. Siminovitch, Genetic and biochemical characterization of mutants of CHO cells resistant to the protein synthesis inhibitor trichodermin, Somat. Cell Genet. 4:355-375 (1978).

200. R.S. Gupta and L. Siminovitch, Diphtheria toxin resistant mutants of CHO cells defective in protein synthesis: A novel phenotype, Somat. Cell Genet. 4:553-571 (1978).

201. J.M. Moehring and T.J. Moehring, Characterization of the diphtheria toxin-resistance system in Chinese hamster ovary cells, Somat. Cell Genet. 5:453-468 (1979).

202. R.S. Gupta and L. Siminovitch, Diphtheria toxin resistance in Chinese hamster cells: Genetic and biochemical characteristics of the mutants affected in protein synthesis, Somat. Cell Genet. 6:361-379 (1980).

203. V. Chan, F.G. Whitmore and L. Siminovitch, Mammalian cells with altered forms of RNA polymerase II, Proc. Natl. Acad. Sci. USA 69:3119-3123 (1972).

204. P.E. Lobban and L. Siminovitch, α-Amanitin resistance: A dominant mutation in CHO cells, Cell 4:167-172 (1975).

205. P.E. Lobban and L. Siminovitch, The RNA polymerase II of an α-amanitin-resistant Chinese hamster ovary cell line, Cell 8:65-70 (1976).

206. C.J. Ingles, A. Guialis, J. Lam and L. Siminovitch, α-amanitin-resistance of RNA polymerase II in mutant Chinese hamster ovary cell lines, J. Biol. Chem. 251:2729-2734 (1976).

207. C.J. Ingles, M.L. Pearson, M. Buchwald, B.C. Beatty, M.M. Crerar, A. Guialis, P.E. Lobban, L. Siminovitch and D.C. Somers, α-Amanitin-resistant mutants of mammalian cells and the regulation of RNA polymerase II activity, in "RNA Polymerase," M. Chamberlin and R. Losick, eds., Cold Spring Harbor Laboratory, Cold Spring Harbor, N.Y., pp. 835-853 (1976).

208. R.S. Gupta, D.H.Y. Chan and L. Siminovitch, Evidence for variation in the number of functional gene copies at the Ama[R] locus in Chinese hamster cell lines, J. Cell Physiol. 97:461-468 (1978).

209. R.S. Gupta and L. Siminovitch, Pactamycin resistance in CHO cells: Morphological changes induced by the drug in wild type and mutant cells, J. Cell Physiol. 102:305-316 (1980).

210. R.S. Gupta, Diphtheria toxin resistance in human lymphoblast lines, Biochem. and Biophys. Res. Commun. 94:1303-1310 (1980).

211. R.S. Gupta and S. Goldstein, Diphtheria toxin resistance in human fibroblast cell strains from normal and cancer-prone individuals, Mutation Res. 73:331-338 (1980).

212. L.B. Jacoby, Canavine-resistant variants of human lymphoblasts, Somat. Cell Genet. 4:221-231 (1978).

213. S.J. Mento and L. Siminovitch, Isolation and preliminary characterization of sindbis virus-resistant Chinese hamster ovary cells, Virology (in press) (1981).

214. P. Coffino, H.R. Bourne, P.A. Insel, K.L. Melmon, G. Johnson and J. Vigne, Studies of cyclic AMP action using mutant tissue culture cells, In Vitro 14:140-145 (1978).

215. T. Haga, E.M. Ross, H.J. Anderson and H.G. Gilman, Adenylate-cyclase uncoupled permanently from hormone receptors in a novel variant of S49 mouse lymphoma cells, Proc. Natl. Acad. Sci. USA 74:2016-2020 (1977).

216. H. Bourne, P. Coffino and G.M. Tomkins, Selection of a variant lymphoma cell deficient in adenylate cyclase, Science 187:750-751 (1975).

217. M.P. Shear, P. Insel, K. Mehnen and P. Coffino, Agonist specific refractoriness induced by isoproterenol, J. Biol. Chem. 251:7572-7576 (1976).

218. P.A. Insel, H.R. Bourne, P. Coffino and G.M. Tomkins, Cyclic AMP-dependent protein kinase-pivotal role in regulation of enzyme-induction and growth, Science 190:896-898 (1975).

219. J. Hochman, P.A. Insel, H.R. Bourne, P. Coffino and G.M. Tomkins, Structural gene mutations affecting regulatory sub-unit of cyclic AMP-dependent protein kinase in mouse lymphoma-cells, Proc. Natl. Acad. Sci. USA 72:5051-5055 (1976).

220. V. Friedrich and P. Coffino, Mutagenesis in S49 mouse lymphoma cells: induction of resistance to ouabain, 6-thioguanine, and dibutyl cyclic AMP, Proc. Natl. Acad. Sci. USA 74:679-683 (1977).

221. R.A. Steinberg, T. van Daalen Wetters and P. Coffino, Kinase negative mutants of S49 mouse lymphoma cells carrying a trans-dominant mutation affecting expression of cAMP-dependent protein kinase, Cell 15:1351-1361 (1978).

222. M.M. Gottesman, A. LeCam, M. Bukowski and I. Pastan, Isolation of multiple classes of mutants of CHO cells resistant to cyclic AMP, Somat. Cell Genet. 6:45-61 (1980).

223. I. Lemaire and P. Coffino, Cyclic AMP-induced cytolysis in S49 cells: Selection of an unresponsive deathless mutant, Cell 2:149-155 (1977).

224. V. Link, J.E. Aubin, A. Chan and F. Sarangi, Mutants of Chinese hamster ovary (CHO) cells with altered colcemid-binding affinity, Cell 18:423-430 (1979).

225. F. Cabral, M.E. Sobel and M.M. Gottesman, CHO mutants resis-
     tant to colchicine, colcemid or griseofulvin have an altered
     β-tubulin, Cell 20:29-36 (1980).

226. R.A.B. Keates, F. Sarangi and V. Link, Structural and func-
     tional alterations in microtubule protein from Chinese hamster
     ovary cell mutants, Proc. Natl. Acad. Sci. USA 78:5638-5642
     (1981).

227. A.E. Lagarde and L. Siminovitch, Studies on Chinese hamster
     ovary mutants showing multiple cross-resistance to oxidative
     phosphorylation inhibitors, Somat. Cell Genet. 5:847-871
     (1979).

228. R.B. Wallace and K.B. Freeman, Selection of mammalian cells
     resistant to a chloramphenicol analog, J. Cell Biol. 65:492-
     498 (1975).

229. T. Lichtor and G.S. Getz, Cytoplasmic inheritance of rutamycin
     resistance in mouse fibroblasts, Proc. Natl. Acad. Sci. USA
     75:324-328 (1978).

230. A. Wiseman and G. Attardi, Cytoplasmically inherited mutations
     of a human cell line resulting in deficient mitochondrial
     protein synthesis, Somat. Cell Genet. 5:241-262 (1979).

231. M. Harris, Cytoplasmic transfer of resistance to antimycin A
     in Chinese hamster cells, Proc. Natl. Acad. Sci. USA 75:5604-
     5608 (1978).

232. M. Rosenstraus and L.A. Chasin, Isolation of mammalian cell
     mutants deficient in glucose-6-phosphate dehydrogenase acti-
     vity: Linkage to hypoxanthine phosphoribosyl transferase,
     Proc. Natl. Acad. Sci. USA 72:493-497 (1975).

233. A.R. Robbins, Isolation of lysosomal α-mannosidase mutants
     of Chinese hamster ovary cells, Proc. Natl. Acad. Sci. USA
     76:1911-1915 (1979).

234. T.D. Stamato and D. Patterson, Biochemical genetic analysis of
     pyrimidine biosynthesis in mammalian cells. II. Isolation and
     characterization of a mutant of Chinese hamster ovary cells
     with defective dihydroorotate dehydrogenase (E.C.1.3.3.1)
     activity, J. Cell Physiol. 98:459-468 (1979).

235. J.D. Esko and C.R.H. Raetz, Replica plating and in situ enzy-
     matic assay of animal cell colonies established on filter
     paper, Proc. Natl. Acad. Sci. USA 75:1190-1193 (1978).

236. J.D. Esko and C.R.H. Raetz, Autoradiographic detection of animal cell membrane mutants altered in phosphatidylcholine synthesis, Proc. Natl. Acad. Sci. USA 77:5192-5196 (1980).

237. T.D. Stamato and C.A. Waldren, Isolation of UV-sensitive variants of CHO-K1 by nylon cloth replica plating, Somat. Cell Genet. 3:431-440 (1977).

238. L.H. Thompson, J.S. Rubin, J.E. Cleaver, G.F. Whitmore and K. Brookman, A screening method for isolating DNA repair-deficient mutants of CHO cells, Somat. Cell Genet. 6:391-405 (1980).

239. D.B. Busch, J.E. Cleaver and D.A. Glaser, Large-scale isolation of UV-sensitive clones of CHO cells, Somat. Cell Genet. 6:407-418 (1980).

240. L. Siminovitch, On the nature of hereditable variation in cultured somatic cells, Cell 7:1-11 (1976).

241. R.S. Gupta, D.Y.H. Chan and L. Siminovitch, Evidence for functional hemizygosity at the Emt locus in CHO cells through segregation analysis, Cell 14:1007-1013 (1978).

242. E.M. Eves and R.A. Farber, Chromosome segregation is frequently associated with the expression of recessive mutations in mouse cells, Proc. Natl. Acad. Sci. USA 78:1768-1772 (1981).

243. C.H. Corsaro and M.L. Pearson, Competence for DNA transfer of ouabain resistance and thymidine kinase: Clonal variation in mouse L-cell recipients, Somat. Cell Genet. 7:617-630 (1981).

CELL HYBRIDIZATION: A TOOL FOR THE STUDY OF CELL DIFFERENTIATION

Mary C. Weiss

Centre de Génétique Moléculaire du C.N.R.S.
91190 Gif-Sur-Yvette, France

INTRODUCTION

One of the many uses of somatic cell hybridization has been to explore the kinds of regulatory mechanisms responsible for the acquisition and the maintenance of the differentiated state. When this work was begun in the late sixties, it was believed that the genetic analysis of cell differentiation, achieved by analyzing the properties of hybrid cells resulting from the fusion of various kinds of differentiated cells, would make it possible to deduce the kinds of genetic mechanisms responsible for the expression of tissue-specific genes. Since that time, it has become clear that regulation in mammalian cells is far too complex to be dissected by this kind of genetic approach alone. Nevertheless, the results obtained do make it possible to eliminate some kinds of models of cell differentiation. In addition, these studies have revealed that fundamentally different mechanisms are involved in the heritable potential of cells to express tissue specific genes, and in the actual expression of these genes. Finally, this line of research has provided the experimenter with the means to alter, in a predictable fashion, the expression of specific genes; this possibility, used in conjunction with the appropriate techniques of molecular biology will make it possible to describe at the molecular level, the changes that are obligatorily associated with the expression or absence of expression of specific genes.

I will give a very brief review of the six major interactions concerning the expression of differentiated functions in hybrid cells; the reader is referred to reviews for documentation (Ephrussi, 1972; Davidson, 1974; Ringertz and Savage, 1976; Weiss, 1977); more recent references will be cited here. Conclusions

concerning our understanding of cell differentiation will be considered, along with problems of interpretation.

Hybridization of somatic cells permits the confrontation of two
genomes in different functional states. From the properties of the
resulting hybrid cells, conclusions can be drawn concerning first,
the apparently "dominant" or "recessive" nature of the expressing
or non-expressing states, and second, the heritability of the commitment of a genome to express specific genes.

Before describing the experimental results obtained from the
study of the expression of differentiated functions in somatic
hybrid cells, it is necessary to present some remarks concerning
the material used and the variables underlying the experiments.
Almost all of the experiments carried out on cell hybrids have been
performed not with normal diploid cells, but with cells of permanent lines, many of them derived from tumors. This choice has been
dictated by technical considerations, among them the ease with
which cells of permanent lines can be manipulated, cloned and used
for the isolation of mutants. Moreover, so far it has proven difficult to obtain pure cultures of normal mammalian cells which
produce tissue-specific proteins in long-term culture in the absence of modification of the karyotype. However, stable, well-
differentiated cell lines can be obtained with relative ease from
tumors which maintain tissue-specific differentiation, and such
tumor-derived cell lines have proved to be amenable to the application of somatic cell genetic techniques such as cell hybridization. It is important to point out that cells of these lines are
not normal, nor are they diploid, so that extrapolation of results
obtained with them to processes of normal development can be made
only with great caution.

Cells of most permanent lines display an aneuploid karyotype,
and it is usual to find some karyotype variation from one cell to
another. The most frequent or modal number of chromosomes per cell
constitutes the stem line of 1s number; it is usually hyperdiploid.
A cell containing twice this number of chromosomes is referred to
as 2s, and the 2s number is usually hypertetraploid.

Tumor-derived differentiated cells representing a number of
tissue types have been used in cell hybridization experiments. We
will consider here only four of them, these being the systems on
which the most extensive analyses have been performed. 1) Mouse
Friend erythroleukemia cells are transformed erythrocyte precursors
that retain the potential to differentiate when stimulated by certain inducers, such as dimethyl sulfoxide. In their growing state,
the tissue-specific proteins characteristic of erythrocytes are not
synthesized; when induced, growth is arrested and synthesis of a
number of erythrocyte specific proteins, including hemoglobin,
commences. When Friend cells are hybridized with other cells, the

response examined is the capacity of hybrid cells to be induced, i.e. to undergo the complex response of Friend cells, including growth arrest and the sequential expression of the functions that characterize erythrocyte differentiation. One of the main advantages of this system is that its biochemistry has been well defined, and probes for globin message have been available for a number of years. 2) Melanoma cells are pigmented at all stages of the growth cycle, although pigment accumulation is most marked in stationary phase. The principal advantage of melanocyte differentiation is that it is visible to the eye; the disadvantage is that its biochemistry is poorly understood. 3) Hepatoma cells produce throughout the course of the growth cycle a number of proteins characteristic of hepatocyte differentiation; the biochemistry is well understood for a number of these functions. Considering these three cell types, it is clear that the question posed in cell hybridization experiments is not the same in each case: for melanoma and hepatoma cells, hybrid cells are examined for the expression of functions that are present at all times in the parent, while for Friend cell hybrids, the inductive response of the parent is a prerequisite for differentiation in the hybrid cells. 4) The last cell type to be considered is the mouse teratocarcinoma cell. Teratoma cells show a striking similarity to early undetermined embryonic cells; they retain the potential, under appropriate conditions, to differentiate into a wide variety of apparently normal non-malignant cell types. With teratoma cell hybrids, one examines the retention or loss of the potential to differentiate. In retrospect it is surprising that, in spite of the differences inherent in the various cell systems utilized, reasonably coherent results have been obtained concerning the experimental situations in which differentiation is and is not expressed in hybrid cells.

We turn now to some of the variables underlying the experiments to be described. When a somatic hybrid is formed, two parental genomes in different functional states can be confronted within the boundaries of a single nucleus. The parental genomes in different functional states fall into two broad categories: those which differ owing to their respective tissue of origin, and those which derive from the same tissue but differ in expression of differentiated functions on account of their developmental stage or through mutation. In the first case, we are dealing with cells of different histogenetic or tissue origin, and therefore of different determination; in the second case, with cells of identical histogenetic origin and identical determination. Two additional variables which must be taken into account in these experiments are the relative ploidy and the species of origin of the parental lines.

The expression of differentiation in hybrid cells can be determined in several types of experimental situations: in the heterokaryon, the immediate product of fusion; in proliferating hybrid cells that retain the essentially complete chromosome complements

of both parents; in hybrid cells that have undergone segregation of
chromosomes; and in the products of fusion of cell fragments (cybrids and reconstituted cells).

Before going on to the results of cell hybridization experiments, it is appropriate to make a few remarks concerning the non-
expressing parent used in crosses. In most of the early work, this
parent was a fibroblast derivative, and this choice was based upon
purely practical considerations: most existing cell lines were of
fibroblast nature, and such cell lines already possessed the selective markers useful for the isolation of hybrid cells. Fibroblasts
are not undifferentiated cells, and for this reason we refer to
expressing and non-expressing, rather than differentiated and undifferentiated parents, and this with reference to a single or a
related group of tissue-specific functions.

THE EXPRESSION OF DIFFERENTIATION IN HYBRID CELLS

In this section, we will define the interactions observed in
hybrid cells, and consider in what experimental situations they
occur.

Co-expression of both parental forms of homologous tissue-
specific functions is observed when the two parents expressed them
prior to fusion. This is most easily demonstrated in interspecific
hybrids where it is possible by serological or physio-chemical
criteria to distinguish the products of both parental genomes
(Cassio and Weiss, 1979; Cassio et al., 1980). It should also be
mentioned that co-expression of both parental forms of metabolic
enzymes is nearly always observed in hybrid cells, and when one
parent is deficient in the activity of such an enzyme owing to
mutation, the homologous gene product of the other parent is expressed.

Extinction of the expression of tissue-specific functions is
generally observed where only one of the parents expressed them
prior to fusion (Ephrussi, 1972). For many of the hybrid types
analyzed, it has been possible to show that no simple trivial explanation can account for absence of expression of the functions:
for example, an inhibitor of an enzyme, difference in species origin of the parental cells, etc. In a few cases analyzed, the absence of expression of a given function is correlated with absence
of the appropriate messenger RNA.

What do we know about the time course of extinction? It occurs
rapidly, and does not require fusion of the parental nuclei
(Mével-Ninio and Weiss, 1981). It may be only transitory in proliferating hybrid cells (Deschatrette et al., 1979; Mével-Ninio and
Weiss, 1981) and transitory extinction can be mediated by the fusion of a cytoplast derived from a cell type known to impose stable

extinction in whole cell hybrids (Kahn et al., 1981).

When the production of several tissue-specific proteins charac-
terizes the expressing parent, the expression of each of these
functions is usually extinguished in hybrid cells. When both parents express differentiated functions but are of widely different
tissue origins, extinction is found to be reciprocal (Fougère and
Weiss, 1978). The extinction of a given function may be only partial. Finally, extinction has been found to occur in hybrids resulting from the two major classes of crosses mentioned above:
those involving cells of different histotypes, and in those resulting from the fusion of cells of identical histotypes, but differing
in the functions expressed (Deschatrette et al., 1979; Cassio and
Weiss, 1979).

From all of these results, the interpretation initially offered
by Davidson, Yamamoto and Ephrussi (see Davidson, 1974) appears to
hold: a cell not expressing a specific function produces diffusible regulatory substances, whose final effects are negative, that
act to prevent expression of the function. We will return to this
point below.

Reexpression of a previously extinguished function can occur.
When first discovered it was observed that reexpression is correlated with massive loss of chromosomes of the non-expressing parent;
from more recent work it appears that in some cases reexpression
may occur with only slight (Bertolotti, 1977) and perhaps only
negligible (Allan and Harrison, 1980; Fougère and Weiss, 1978)
chromosome loss. There are no confirmed reports of the association
of loss of an identified chromosome with the reexpression of a
given function. When the expressing parent produced a number of
tissue-specific proteins, reexpression of the individual functions
was found to occur independently and in no apparent order.

These observations demonstrate that extinction does not correspond to a loss of determination of the genome of the expressing
parent: extinction causes merely a block to expression of the
function analyzed, but does not modify the potential of the genome
to reestablish its original functional state. The fact that reexpression of individual functions occurs independently is most
coherent with the idea that each function is extinguished by an
independent mechanism. However, if a two-step mechanism is in-
voked, this interpretation can be re-considered (see below).

Gene dosage effect describes the observation that doubling of
the ploidy of the expressing parent leads to the formation of hybrids in which extinction is not observed. For example, when 1s
melanoma cells are fused with 1s fibroblasts, extinction of pigment
production is systematic; 2s melanoma cells crossed with the same
1s fibroblasts leads to the formation of hybrid cells that

synthesize large amounts of melanin. These and other observations
of gene dosage effects demonstrate that it is possible for the
entire pattern of gene expression of the hybrid cells to swing one
way or the other: when extinction is observed it is often complete
(within the limits of sensitivity of the detection methods); when
extinction does not occur, owing to gene dosage effects, the hybrid
cells produce large amounts of the tissue-specific proteins of the
expressing parent. It has also been observed that maintenance of
the "differentiated" phenotype of such hybrids is dependent upon
the retention of large numbers of chromosomes of the expressing
parent, and that "absence of extinction" can correspond to reexpression, following a transitory extinction whose time course is
identical to that observed in 1s hybrids (Mevel-Ninio and Weiss,
1981). A pleiotropic modulation of the hybrid cell phenotype,
similar to that observed for gene dosage effects, has been reported
for Friend cell-fibroblast hybrids, and this in the absence of any
apparent karyotype differences: hybrid cells that remain attached
to the substrate, like the fibroblast parent, fail to show inducibility of globin synthesis; those that grow in suspension like the
Friend cell parent, do show inducibility of globin production
(Axelrod et al., 1978; Allan and Harrison, 1980). All of these
observations, unlike those reported above concerning reexpression,
lend support to the idea that numerous factors influence (in a
specific or in a relatively non-specific fashion) the final expression of a number of tissue-specific genes. Moreover, one is led to
postulate that the final balance of these factors (contributed
presumably by both parental genomes) determines in a highly pleiotropic way the phenotype of the hybrid cells: morphology, a para-
meter that cannot be measured but clearly reflects the expression
of many genes, is surprisingly strictly correlated with the expression of a few tissue-specific functions, parameters that can be
measured and are consequently emphasized.

Activation is often observed when extinction does not occur or
is only partial, and corresponds to the synthesis, directed by the
non-expressing genome, of its homologue of a tissue-specific protein. Activation reflects the induced activity of a previously
silent gene, and most likely corresponds to a transcriptional activation combined with all subsequent steps necessary for the final
expression of a gene. It demonstrates that structural genes whose
expression should never be observed, have not been modified during
the course of development in a fashion that would render them re-
fractory to activation. Activation is systematic in crosses where
gene dosage effects are in favor of the expressing parent; it appears to hold for multiple functions of a given type of differentiation; it may occur even when there is partial (up to 90%) extinction of the expression of a given function; and it may accompany
reexpression (Bertolotti, 1977). It remains unknown whether activation occurs simply because an extinction mechanism, operative in

the non-expressing parental cell, is effectively diluted in the hybrid, or whether a discrete activating substance that is active upon both parental genomes, is contributed by the expressing parent. It is clear that activation does not reflect a modification of the determination of the "non-expressing" genome. Rat hepatoma-mouse melanoma hybrids show reciprocal extinction of albumin and pigment production; hybrid subclones that show reexpression of rat albumin and activation of mouse albumin synthesis retain the potential to synthesize pigment. Thus, it is possible to obtain synthesis of a tissue-specific protein (albumin) foreign to the determination of the melanoma parental genome without its "forgetting" its original determination as a melanocyte (Fougère and Weiss, 1978).

Retention of pluripotency has been observed in some types of teratoma cell hybrids. Pluripotent teratoma cells, fused with mouse L fibroblast cells, give rise to hybrids that resemble the L cell parent. However, when fused with diploid thymocytes or pseudo-diploid Friend cells, teratoma hybrids may resemble their pluripotent teratoma parent (Miller and Ruddle, 1976 and 1977); moreover, some teratoma-Friend cell hybrids show the phenotype of the Friend parent, and the globin genes of both are expressed (McBurney et al., 1978). We see again here cases where the hybrid cell phenotype follows in a highly pleiotropic manner that of one or the other parent. It is perhaps surprising to note that it appears to be possible for the genome of a differentiated somatic cell to be re-programmed to an undetermined pluripotent state. It is only in the case of these teratocarcinoma hybrids that evidence has been obtained of a modification of determination ensuing from the hybridization of somatic cells.

Phenotypic exclusion has been observed in hybrids obtained by crossing cells of widely different histotypes, but not in those whose parents were of very similar lineage. Thus, in melanoma (neural crest)-hepatoma (endoderm) hybrids, albumin production and pigment synthesis were never observed to be reexpressed simultaneously (Fougère and Weiss, 1978). By contrast hybrids between Friend cells and lymphoma or myeloma cells were found to produce the surface antigens characteristic of the lymphoma or myeloma parent, and at the same time to be inducible for the synthesis of hemoglobin (Allan and Harrison, 1980). More experiments will clearly have to be done before conclusions can be drawn concerning the generality of phenotypic exclusion or its absence, and whether or not the results are a reflection of the relatedness of the embryonic lineage of the parental cells, or due to totally unrelated parameters.

PROBLEMS OF INTERPRETATION

The observations concerning extinction, reexpression, activa-

tion, and phenotypic exclusion are most compatible with the idea
that highly specific regulatory substances are involved in
determining whether or not individual tissue-specific functions
will be expressed in hybrid cells. By contrast, the results obtained on gene dosage effects, the retention of pluripotency, and
correlations between hybrid cell morphology and the expression of
hemoglobin inducibility would seem to argue that multiple factors
may determine in a highly pleiotropic fashion the differentiation
competence of hybrid cells, and that this may be unrelated to
direct regulation of expression of the tissue specific genes, acting via, for example, the expression of proteins that determine
cell architecture and the cytoskeleton (see Allan and Harrison,
1980). While these interpretations may appear contradictory, it is
worthwhile considering the nature of the differences inherent in
the parental cells used. When one is dealing with cells that express their differentiated functions at all times, such as melanoma
or hepatoma cells, it would not be surprising to observe in hybrid
cells a direct regulation of the expression of tissue-specific
genes. By contrast, when we consider cell types that retain the
potential, under appropriate environmental conditions to carry out
a specific differentiation program which is normally expressed in a
sequential manner, it would not be surprising to find that there is
an obligatory order to expression of the functions, and that a
block in an early step (perhaps related to cell architecture) is
sufficient to inhibit subsequent steps.

With the exception of crosses concerning teratoma cells, in no
case has it been possible to obtain evidence in favor of a heritable modification of the determination of one parent. A number of
investigators have attempted to search directly for such a heritable modification, using the techniques of fusion of cell fragments. Contradictory results have been obtained. For example,
several investigators (Ringertz et al., 1978; Linder et al., 1979;
McBurney and Strutt, 1979) have failed to obtain a heritable effect
when myoblast karyoplasts or teratoma cells are fused with cytoplasts of a cell of a different species that would cause extinction
in whole cell fusions (see also Kahn et al., 1981). In contrast,
Gopalakrishnan et al. (1977) have reported cases of the stable
extinction of hemoglobin inducibility in cybrids between Friend
cells and cytoplasts. By contrast, Lipsch et al., (1979) have
reported that fusion of L cell karyoplasts with rat hepatoma cytoplasts leads to the expression of the mouse form of an enzyme characteristic of the hepatoma line, and that this expression is stable
during many tens of generations, and then is abruptly lost. The
most striking result has been obtained by Gopalakrishnan and
Anderson (1979); they found that fusion of mouse Friend cells with
rat hepatoma cytoplasts can lead to the expression of the mouse
form of a liver specific enzyme whose production is maintained as
long as there is selective pressure for its presence. The simplest
interpretation of the examples where a positive effect has been

obtained is that a regulatory factor, that causes extinction or activation, can be transmitted via a cytoplast and that the further production of this factor by the resident genome reflects autoregulation. For the moment, too few examples of a heritable effect have been reported for any conclusion to warrant generalization.

CONCLUSIONS

Extinction, dosage effects and activation provide insight into the problem of the regulation of overt expression of differentiated functions, where the final effect is either negative (extinction) or positive (activation). Since these interactions occur when two genomes in different functional states are simply confronted, we are forced to conclude that diffusible substances are involved, although we have as yet no idea of their chemical nature nor of their sites of action. In the case of activation however, it appears highly probable that the substance is acting at the level of the responding gene, for example involving reorganization of the chromatin.

Reexpression of previously extinguished functions by segregated hybrid cells demonstrates that the determination of somatic cells can be inherited in the absence of overt expression. We must therefore conclude that determination is due to an autonomous mechanism that is not altered or lost in cell hybrids. However, we have already seen above that diffusible substances can act in cell hybrids to alter drastically the pattern of proteins that is actually produced, including cases of activation of a genome to produce a protein foreign to its own determination. Nevertheless, both parental determinations coexist in the hybrid nucleus, in spite of the fact that one genome has been activated to participate in the synthesis of a protein foreign to its own determination.

From these various observations of interactions in hybrid cells, it appears clear that there are at least two fundamentally different levels of regulation in mammalian somatic cells, that concerned with determination or genomic commitment, and that responsible for expression of tissue-specific proteins. The former is characterized by an extraordinary degree of stability and is not lost when cells of different determinations are hybridized; the latter appears to be regulated by trans-acting substances that can have a final effect that is negative (in the case of extinction) or positive (activation), without causing loss of the original determination of the cell's genome. In contrast to these conclusions, which appear to apply to all cases examined that involve cells of a somatic derivative (i.e. differentiated cells), some somatic cells appear to lose their determination when fused with pluripotent teratoma cells.

THE OPEN QUESTION OF MECHANISMS

   For none of the interactions described is a mechanism known.
This, to me, is not surprising or even disappointing. To understand a mechanism, it is not sufficient to determine the molecular
level at which something happens: one must know not only "what"
makes it happen, but also how that "what" is regulated. When these
questions are answered, be it concerning events that occur in
somatic hybrids or in any other experimental system appropriate for
the study of differentiation, we will have learned a great deal.
In any case, it does not appear likely that these questions can be
answered in the absence of a "genetic" test, for this is the only
means to determine the functional role of a molecule.

   Before continuing with a discussion of types of mechanisms
compatible with the observations described, it appears appropriate
to ask whether this line of research merits continuation. Are we
dealing with 1) totally non-specific effects; 2) specific effects
that are relevant primarily to an understanding of the behavior of
malignant cells of permanent lines; 3) or specific effects that
reveal how gene expression is regulated during development. To me,
the answer lies somewhere between the second and third possibilities. There are a number of arguments that in at least some experimental situations the third possibility holds. First, activation
is a highly specific effect: a previously unexpressed gene may
become active, and this is always a gene that is expressed (or is
potentially expressable) by the activating parent (Willing et al.,
1979). Continuation of the study of activation will make it possible to define at the level of the gene itself the changes that
occur when a gene shifts from a non-expressing to an expressing
state. In addition, it may permit identification of the substance
responsible for activation. The phenomenon of extinction may be
caused by more than one mechanism (see Mével-Ninio and Weiss,
1981). However, it may occur in hybrids formed by crossing cells
of the same histotype (Cassio and Weiss, 1979; Deschatrette et al.,
1979; Forquignon and Ephrussi, 1979), which suggests that even a
differentiated cell which fails to express one or more of the functions characteristic of its own differentiation uses a transacting
mechanism to prevent expression of the gene. In addition, our
recent results (Cassio et al., Sala-Trépat et al., manuscript submitted) show that the level of the block to the expression of the
albumin gene appears to be same in albumin-negative rat hepatoma
variants, and in their albumin-producing cells of origin. A search
for the factor(s) causing extinction does indeed seem merited,
particularly if the system chosen involves cells of not too distinct tissue origin. Still another aspect of research involving
hybrid cells seems indicated: for an understanding of normal development, hybrid cells can be used to verify any indication that a
molecular footprint of determination has been found; this can be
tested upon hybrids where the parental genomes do retain their

original determination in the absence of expression of differentiated functions, and in teratoma hybrids where the differentiated
parental genome appears to have lost its determination.

We turn now to the mechanisms underlying the interactions observed. The reader will have noted here, and in original publications in the field, that use is not made of the words "repression"
and "induction". This is because these words connote specific
mechanisms. The results available do not even make it possible to
reason in terms of known procaryotic models of regulation (see the
discussion in Davis and Adelberg, 1973); for example, a final negative effect may be due to the action of a repressor like molecule,
or an activator of a repressor, etc. In no case do we know whether
a one-step or a multi-step mechanism is responsible for the final
effect. Moreover, there is no a priori reason to limit the possibilities to known procaryotic models.

Why is it that we do not even know whether a one-step or a multi-
step process is involved in the various interactions? Take the
case of extinction, an event that occurs rapidly and does not even
require the fusion of the parental nuclei; we will consider two
simple, but very different, kinds of explanations, and both are
compatible with the known facts. 1) Extinction is caused by a
diffusible regulatory molecule produced by the non-expressing
parent that acts directly at the level of the appropriate structural gene of the expressing parent. 2) The non-expressing parent
produces a factor that activates the genome of the expressing
parent to produce the very molecule it uses to inhibit the expression of its own differentiation at an earlier developmental stage.
If we can't even say with assurance which genome is responsible for
the production of a factor that acts on the appropriate gene, model
building clearly seems premature.

As mentioned above, two kinds of explanations are compatible
with the observations of reexpression, and these apply to activation as well. 1) A chromosome of the non-expressing parent, that
codes for the factor causing extinction of the function in question, is lost, and this loss alone is sufficient to permit reexpression and even activation. 2) The occurrence of reexpression
and of activation requires a critical quantity of an activating
factor synthesized by the expressing genome.

These very fundamental questions will be answered in an unam-
biguous fashion only when the molecules responsible for the effects
have been identified. This identification should be feasible in
view of the variety of powerful techniques currently available.

A discussion of the mechanisms involved in the highly pleiotropic changes described above does not appear useful at this
stage, for the possibilities are too numerous.

We turn finally to the question of the striking heritability of determination. In differentiated somatic cells, genomic commitment is not influenced by the factors that modify the final expression of genes coding for tissue-specific proteins. In addition, there are no compelling arguments that the expression of these very factors is modified in hybrid cells: all of the observations concerning the expression of tissue-specific proteins are compatible with the notion that both parental genomes continue, after fusion, to synthesize the same regulatory factors that they produced before fusion, and that the final phenotype of the hybrid is a reflection of the balance of these factors.

In conclusion, the work described has provided some clear answers, it has made us pose many questions, and provides a complex but solid framework for the interpretation of future experiments.

ACKNOWLEDGEMENTS

Numerous discussions with present and past colleagues in the laboratory have led to the conclusions presented here. Linda Sperling kindly read the manuscript.

REFERENCES

Allan, M. and Harrison, P., 1980, Co-expression of differentiation markers in hybrids between Friend cells and lymploid cells and the influence of cell shape. Cell 19:437-447.

Axelrod, D.E., Gopalakrishnan, T.V., Willing, M. and Anderson, W.F., 1978, Maintenance of hemoglobin inducibility in somatic cell hybrids of tetraploid (2s) mouse erythroleukemia cells with mouse or human fibroblasts. Somat. Cell Genet. 4:157-168.

Bertolotti, R., 1977, Expression of differentiated functions in hepatoma cell hybrids: selection in glucose-free media of hybrid cells which reexpress gluconeogenic enzymes. Somat. Cell Genet. 3:579-602.

Cassio, D., Hassoux, R., Dupiers, M., Uriel, J. and Weiss, M.C., 1980, Coordinate secretion of mouse alpha-fetoprotein, mouse albumin and rat albumin in mouse hepatoma-rat hepatoma hybrids. J. Cell Physiol. 104:295-308.

Cassio, D. and Weiss, M.C., 1979, Expression of fetal and neonatal hepatic functions by mouse hepatoma-rat hepatoma hybrids. Somat. Cell Genet. 5:719-738.

Davidson, R.L., 1974, Gene expression in somatic cell hybrids. Annu. Rev. Genetics 8:195-218.

Davis, F.M. and Adelberg, A., 1973, Use of somatic cell hybrids for analysis of the differentiated state. Bacteriol. Rev. 27:197-214.

Deschatrette, J., Moore, E.E., Dubois, M., Cassio, D. and Weiss, M.C., 1979, Dedifferentiated variants of a rat hepatoma: analysis by cell hybridization. Somat. Cell Genet. 5:697-718.

Ephrussi, B., 1972, "Hybridization of somatic cells". Princeton University Press, Princeton, NJ.

Forquignon, F. and Ephrussi, B., 1979, Isolation and properties of amelanotic variants of a hamster melanoma. Somat. Cell Genet. 5:409-426.

Fougère, C. and Weiss, M.C., 1978, Phenotypic exclusion in mouse melanoma-rat hepatoma hybrid cells: pigment and albumin production are not reexpressed simultaneously. Cell 15:843-854.

Gopalakrishnan, T.V. and Anderson, W.F., 1979, Epigenetic activation of phenylalanine hydroxylase in mouse erythroleukemia cells by the cytoplast of rat hepatoma cells. Proc. Natl. Acad. Sci. USA 76:3932-3936.

Gopalakrishnan, T.V., Thompson, E.B. and Anderson, W.F., 1977, Extinction of hemoglobin inducibility in Friend erythroleukemia cells by fusion with cytoplasm of enucleated mouse neuroblastoma or fibroblast cells. Proc. Natl. Acad. Sci. USA 74:1642-1646.

Kahn, C.R., Bertolotti, R., Ninio, M. and Weiss, M.C., 1981, Short-lived cytoplasmic regulators of gene expression in cell cybrids. Nature 290:717-720.

Linder, S., Brzeski, H. and Ringertz, N.R., 1979, Phenotypic expression in cybrids derived from teratocarcinoma cells fused with myoblast cytoplasms. Exp. Cell Res. 120:1-14.

Lipsch, L.A., Kates, J.R. and Lucus, J.J., 1979, Expression of a liver-specific function by mouse fibroblast nuclei transplanted into rat hepatoma cytoplasts. Nature 281:74-76.

McBurney, M.W., Featherstone, M.S. and Kaplan, H., 1978, Activation of teratocarcinoma-derived hemoglobin genes in teratocarcinoma-Friend cell hybrids. Cell 15:1323-1330.

McBurney, M.W. and Strutt, B., 1979, Fusion of embryonal carcinoma cells to fibroblast cells, cytoplasts and karyoplasts: developmental properties of viable fusion products. Exp. Cell Res. 124:171-180.

Mével-Ninio, M. and Weiss, M.C., 1981, Immunofluorescent analysis
of the time course of extinction, reexpression and activation of
albumin production in heterokaryons and hybrids of rat hepatoma
cells with mouse fibroblasts. J. Cell Biol., in press.

Miller, R.A. and Ruddle, F.H., 1976, Pluripotent teratocarcinoma
thymus somatic cell hybrids. Cell 9:45-55.

Miller, R.A. and Ruddle, F.H., 1977, Teratocarcinoma x Friend
erythroleukemia cell hybrids resemble their pluripotent embryonal
carcinoma parent. Dev. Biol. 56:157-173.

Ringertz, N.R., Krondahl, V. and Coleman, J.R., 1978, Reconstitu-
tion of cells by fusion of cell fragments. I. Myogenic expression
after fusion of minicells from rat myoblasts (L6) with mouse fibroblast (A9) cytoplasm. Exp. Cell Res. 113:233-246.

Ringertz, N. and Savage, R.E., 1976, "Cell hybrids". Academic
Press, New York.

Weiss, M.C., 1977, The use of somatic cell hybridization to probe
the mechanisms which maintain cell differentiation. In "Human
Genetics", A. Armendares and R. Liskey, eds., Excerpta Medica,
Amsterdam, 284-292.

Willing, M.C., Nienhuis, A.W. and Anderson, W.F., 1979, Selective
activation of human β- but not γ-globin gene in human fibroblast x mouse erythroleukemia cell hybrids. Nature 277:534-538.

USE OF SOMATIC CELL HYBRIDIZATION AND DNA-MEDIATED GENE TRANSFER FOR CHARACTERIZATION OF NEOPLASTIC TRANSFORMATION

Klaus Willecke and Reinhold Schäfer

Institut für Zellbiologie (Tumorforschung)
Universität Essen, Hufelandstr. 55, 4300 Essen 1
Federal Republic of Germany

INTRODUCTION

Somatic cell hybridization has been used with great success for the assignment of genes to human or mouse chromosomes (1). In most cases, phenotypes have been correlated with <u>single</u> genes on chromosomes. Complex phenotypes which are caused by the expression of several <u>different</u> genes are difficult to dissect genetically by the standard methods of somatic cell genetics without additional use of techniques of molecular biology. It appears that neoplastic transformation is such a complex phenotype. The expression of malignancy in somatic cell hybrids has been investigated already in one of the erliest publications describing the cell fusion phenomenon (2). Since then many different cell systems and expression of many different phenotypes of neoplastic transformation have been studied in somatic cell hybrids. In particular, the question of dominant versus recessive expression of malignancy in cell hybrids has led to many controversial discussions in the literature. Very recently, the technique of DNA-mediated gene transfer has been successfully applied to the transfer of genes (or DNA sequences) involved in tumorigenesis. It appears that further exploration of this technique in combination with molecular gene cloning could lead to deciphering some of the puzzling effects regarding expression of malignancy in somatic cell hybrids.

In this chapter, we wish to summarize the present knowledge of this research field and to draw several general conclusions, some of them being still preliminary. Because of limits of space, we do not discuss all papers dealing with a certain question.

The relevant literature of somatic cell hybridization has been
extensively reviewed (3-7). Here we present a synopsis of the
research fields of somatic cell genetics and DNA-mediated gene
transfer with regard to expression of malignancy.

1. DURING TRANSITION OF A NORMAL CELL TO A TUMORIGENIC CELL EXPRESSION OF MANY GENES IS CHANGED

There exists an extensive literature describing changes in
the expression of phenotypes during cellular transformation. Many
of these phenotypes have been studied in somatic cell hybrids,
for example permanent cell proliferation, changes in cytoskeleton,
membrane glycoproteins, extracellular matrix (i.e. fibronectin),
and plasminogen activator, loss of density dependent growth control, and independence of serum growth factors and of anchorage.
For most of these phenotypes, exceptional cells or cell lines
have been described, which can form tumors and yet lack the
"typical" change of a certain phenotype found in most other tumorigenic cells. The existence of these exceptional cells could
point to different molecular mechanisms of tumorigenicity. More
recently, changes in the fucosyl glycopeptides (8) and in a cell
surface glycoprotein (9) have been found between normal and transformed cells, so far without exception. However, more independent
cell lines have to be assayed before those criteria can be
accepted as being related to a common mechanism of malignancy or ,
for example, to cell proliferation (10).

By studying phenotypes connected with neoplastic transformation in somatic cell hybrids one had hoped to define and
possibly map genes involved in expression of malignancy. For
example, human genes coding for plasminogen activator and fibronectin have been assigned to specific chromosomes (11,12). However, it has not yet been possible to map genes involved in
expression of more complex phenotypes of neoplastic transformation, such as independence of anchorage.

2. SOMATIC CELL HYBRIDS OF TWO DIFFERENT TUMOR CELL LINES ARE USUALLY TUMORIGENIC

A variety of somatic cell hybrids between different mouse
cell lines, transformed by carcinogens or viruses, gave the same
tumor incidences as the corresponding parental cells. Occassional-
ly malignancy of hybrids was reduced to the tumor incidence of
the less malignant parent (3). The absence of complementation
between different tumor cells could be explained by the dominant
expression of the malignant phenotype. However, it cannot be
excluded that these malignant cell lines commonly underwent
mutations at the same genetic locus and thus failed to complement
each other in somatic cell hybrids (5). Recently, complementation

of malignancy in somatic cell hybrids between mouse tumor cells
and Syrian hamster tumor cells was reported (13), indicating the
recessive character of malignant changes in these cells.

3. SOMATIC CELL HYBRIDS OF TUMORIGENIC AND NORMAL CELLS ARE
    FREQUENTLY NON-TUMORIGENIC

The concept that malignancy is due to recessive changes in
the genome of a neoplastic cell stems from results obtained by
analyzing hybrids between tumorigenic cells and normal cells.
Intraspecific mouse, human, and hamster hybrids, and interspecific
human x mouse (tumorigenic cell line indicated first), hamster x
mouse, mouse x human, hamster x human, and mouse x rat somatic
cell hybrids were shown to behave like the non-tumorigenic,
diploid parent after injection into susceptible hosts (suppression
of malignancy). Contrary to this, some authors concluded from
their results that malignancy is a dominant trait, since they
observed expression of tumorigenicity in intraspecific mouse or
human hybrids and in interspecific human x mouse, mouse x human,
and mouse x hamster hybrids (cf.3-7). These controversial conclusions may, at least in part, be due to different experimental
approaches, such as the use of parental cells of different tissue
origin and different immunosuppressed animals for assessment of
tumor growth, or due to an insufficient quantitation of the
tumorigenic potential of hybrid cells. In some hybrid systems,
the strong selective pressure against non-tumorigenic cells, once
the hybrid cells had been injected into the host, produced tumorigenic segregants at such a high frequency that suppression of
malignancy could have been obscured (5). In particular,
interspecific hybrids show a very rapid segregation of chromosomes
of the normal parent.

Contradictory results have also been obtained with regard to
growth in semi-solid agar of hybrids between anchorage-independent
tumor cell lines and normal cells (for a summary see 14). Intra-
and interspecific hybrids were interpreted to show either dominant
expression or suppression of growth in semi-solid medium.
Anchorage independence and tumorigenicity were shown to be under
separate genetic control in intraspecific human hybrids (15,16).

All somatic cell hybrids involving permanent cell lines and
normal cells exhibited unlimited proliferation in vitro (escape
from senescence). It has been reported, however, that senescent
human diploid fibroblasts after fusion with karyoplasts or whole
cells derived from permanent lines show a finite life-span (17,
18). Obviously, senescence behaved in a dominant manner in these
cell hybrids. This phenomenon could not be observed in most of
the fusions concerning the expression of malignancy, since permanent proliferation of the somatic cell hybrids to be isolated
had been used as a selective marker.

4.   SEGREGATION OF CHROMOSOMES FROM NON-TUMORIGENIC HYBRIDS
     COINCIDES WITH REEXPRESSION OF MALIGNANCY

     The injection into susceptible hosts of mass populations of
somatic cell hybrids (10^5 to 10^7 cells), which were non-tumorigenic at lower cell dose, resulted in a more or less frequent
occurence of tumors (reexpression of malignancy). These tumor
hybrids were subpopulations derived from the original hybrid isolates, since they had lost chromosomes of the normal parent (5,
7). Assuming that tumorigenicity in somatic cell hybrids is
recessive and normal cells carry gene(s) capable of suppression
of malignancy, several authors tried to map these genes by com-
paring the chromosome content of suppressed hybrids and of tumorigenic segregants derived from them. In particular, tumorigenic
segregants were assessed for any consistent loss of a specific
chromosome of the normal parent. Mouse chromosome 4 was re-
gularly lost in tumors derived from hybrids involving normal
mouse cells and a mouse melanoma line (19). This result, however,
was not confirmed by other investigators, since mouse chromosome 4
was frequently retained in mouse tumor x mouse fibroblast hybrids
obtained in vivo and in vitro (20,21). Obviously, suppression of
malignancy is not due to the action of a single gene, since no
consistent loss of any single chromosome of the normal parent was
found in tumorigenic interspecific mouse x human hybrids (22),
hamster x mouse hybrids (14), and intraspecific human hybrids
(16). In these human hybrids, certain combinations of two human
chromosomes were frequently absent. Therefore, genes on at least
two different human chromosomes appeared to be required for
suppression (16). Tumorigenic intraspecific mouse hybrids showed
no loss of any specific chromosome pair contributed by the normal
parent (21). All combinations of two mouse chromosomes were found
in Chinese hamster x mouse hybrid tumors except three combinations
involving mouse chromosome 11 (14). These data suggest that more
than two genes located on different mouse chromosomes may have to
be expressed simultaneously to suppress malignancy. Furthermore,
two copies of certain mouse chromosomes, or pairs of mouse chromosomes, respectively, were only rarely retained in these tumor
hybrids. Therefore, the presence of multiple copies of normal
mouse chromosomes may be a prerequisite for non-expression of
malignancy of Chinese hamster x mouse somatic cell hybrids in
nude mice (14, see also paragraph 5).

     The contribution of extrachromosomal genes in expression of
the transformed state has also been examined in different hybrid
systems. Mouse x human hybrids were non-tumorigenic despite the
total loss of cytologically detectable human chromosomes (23).
Cybrids involving tumorigenic Chinese hamster cells and cytoplasts

derived from non-tumorigenic Chinese hamster cells showed
suppression of malignancy (24). Thus, it appears that extrachromosomal elements or cytoplasmic factors can contribute to
suppression of malignancy.

5. EXPRESSION OF TUMORIGENICITY IN SOMATIC CELL HYBRIDS APPEARS
TO BE DEPENDENT ON THE DOSAGE OF THE PARENTAL GENOMES

Somatic cell hybrids between Chinese hamster cells and normal mouse fibroblasts were as tumorigenic in nude mice as the
hamster parental cells when they contained 50 to 66 Chinese
hamster chromosomes (tetraploid number: 44; hexaploid number: 66)
and 1 to 18 mouse chromosomes (14). One exceptional hybrid required 1,000 fold more cells to initiate tumor growth (minimal
cell number 5×10^4) than the hamster parent (minimal cell number
50, coinjected together with 5×10^6 X-irradiated mouse fibroblasts according to (25). This hybrid contained 47 hamster chromosomes and 17 mouse chromosomes (mean chromosome number). On
the other hand, hybrids containing an equal number of mouse
chromosomes, but 55 to 57 hamster chromosomes did not show any
reduction in tumor-forming ability. In order to find out,
whether suppression of malignancy in the hamster x mouse hybrid
system is dependent on the dosage of the two parental genomes,
we produced hybrids between a near diploid derivative of the
Chinese hamster cells and mouse fibroblasts (20 to 50 fold
excess). Three suppressed hybrids contained Chinese hamster
chromosomes at a near diploid, a near triploid and a near tetraploid level together with 22 to 30 mouse chromosomes, representing all the different chromosomes of a normal mouse genome.*
These hybrids required 10^2 to 10^4 more cells to initiate tumor
growth in nude mice and showed a three-to sixfold increase in
latency period. We assume that the gene alterations responsible
for expression of malignancy in the Chinese hamster genome can
exist in multiple copies. Tumorigenicity can be suppressed in
hybrids only, if the Chinese hamster genome consists of a near
diploid chromosome set or if the gene dosage of the normal
mouse genome is increased above a critical number of mouse chromosomes.

In Syrian hamster cells (26) and in hybrids between tumorigenic mouse cells and normal rat cells (13), it has been documen-
ted that putative "tumor genes" cannot be suppressed if present
in multiple copies. Our data support the hypothesis that the
expression of malignancy can be controlled by the dosage of normal genes and putative "tumor genes".

---

*Unpublished results

6.   SOMATIC CELL HYBRIDS OF CERTAIN TUMOR CELLS WITH NORMAL
     CELLS APPEAR TO BE ALWAYS TUMORIGENIC

     Hybrids between mouse myeloma cells and normal B lymphocytes
(hybridomas) were tumorigenic in syngeneic mice or nude mice after
injection of 10^6-10^7 cells (7).   However, evidence for tumor
growth after inoculation of lower cell numbers was not reported.
No preferential loss of specific chromosomes of the normal
parent was detected.   Hybrids between human fibrosarcoma and
osteosarcoma cells and different human fibroblasts were reported
to be always tumorigenic (at 5×10^6 injected cells per nude mouse)
(7).   These results are in contrast to those described for Hela x
human fibroblast hybrids, in which a very stable suppression was
observed (15).   The human fibrosarcoma x fibroblast hybrid system
will be useful to verify the hypothesis that the tumorigenic
phenotype can be transferred to a normal genome in a dominant
fashion, either by microcell fusion or DNA-mediated gene transfer.

7.   TRANSFER OF DNA FROM CERTAIN CHEMICALLY TRANSFORMED CELL
     LINES INTO PRENEOPLASTIC MOUSE 3T3 CELLS LEADS TO NEOPLASTIC
     TRANSFORMATION

     In 1979 it was reported (27) that high molecular weight DNA
or chromatin from chemically transformed mouse or rat cells could
be used to transfect mouse NIH3T3 cells monitored by formation of
foci on a confluent cell layer.   (The term "transfection" had been
originally suggested for the detection of biological activity of
viral DNA extracted from transformed cells (28,29)).   Furthermore,
the authors (27) reported that cells from these foci had gained
the ability to grow in soft agar and were tumorigenic in newborn
mice.   Thus, it appears that the phenotype of neoplastic transformation can be transferred into preneoplastic mouse cells.
Using the same experimental system other workers (30)
independently confirmed and extended the results described in
(27).   More recently, both groups have shown that DNA from certain
human bladder carcinoma cell lines can be successfully used in
the transfection assay (31,32).   Furthermore, it was found (31)
that DNA obtained from rabbit and mouse carcinoma and from rat
neuroblastoma cell lines was able to induce neoplastic transformation of NIH3T3 cells.   Several important conclusions can be
drawn from these results:

(i) It appears that certain tumorigenic cells contain a DNA
sequence which after transfer into 3T3 cells causes dominant
expression of neoplastic transformation.   So far, successful experiments of this type have been described only for 3T3 recipient
cells.   3T3 cells exhibit density dependent growth control and

do not form tumors in appropriate mice, similar to normal cells. However, in contrast to normal cells, 3T3 cells are a permanently growing mouse cell line containing an abnormal set of chromosomes. 3T3 cells attached to plastic discs are tumorigenic after implantation into mice. Thus, 3T3 cells are termed preneoplastic. It is possible that normal mouse cells, when used for transfection experiments with DNA from certain chemically transformed cell lines, may not express the transformed phenotypes.

(ii) It is unlikely that viral DNAs integrated in certain tumorigenic cellular DNAs were responsible for transfection of 3T3 cells. All mentioned reports list appropriate controls. In none of the reported transfection experiments had any release of transforming virus been found.

(iii) Presumably, the induction of neoplastic transformation by transfection of 3T3 cells is due to the transfer of a single DNA fragment of less than 30 kilobase pairs size. This can be concluded from the relatively low efficiency of transformation and the dilution of cellular DNA used for the transfection experiments.

(iv) Successful transfection with purified chromosomes from two different tumorigenic donor cell lines was reported (27). Using the same donor cells, in one of these cases DNA mediated transfection has been unsuccessful. Further investigation has to show whether additional DNA introduced with chromosomes is required for successful transfection.

(v) All authors reported that transfection of 3T3 cells was successful only with DNA from relatively few of the tumorigenic cell lines which were assayed. This result points to the heterogeneity of tumorigenic cell lines. So far, no positive results with DNA from primary human tumors have been obtained (32). On the other hand, successful transfections have been carried out with DNA derived from tumorigenic cell lines of several tissues and species (mouse, rat, and human).

(vi) Highly repetetive human DNA ("Alu-family") was found in only those 3T3-derived clones which had been obtained after transfection with DNA from human bladder cell carcinoma cell lines (31,32). Both groups used cloned human Alu DNA-sequences as a hybridization probe (33). These sequences did not hybridize with DNA from preneoplastic mouse 3T3 cells. These results prove that human DNA has indeed been transferred into "bona fide" transfectants. Thus presumably, neoplastic transformation of these 3T3-derived clones was due to transfection by human DNA rather than due to any spontaneous event.

8. THE DNA SEQUENCES OF DIFFERENT TUMORIGENIC CELL LINES WHICH
    GIVE RISE TO TRANSFECTIONS OF NEOPLASTIC TRANSFORMATION MAY
    BE DIFFERENT FROM EACH OTHER

    Up to now, none of the transfecting DNA sequences from
chemically transformed cell lines have been molecularly cloned in
bacteria. On the other hand, several transforming genes of murine
retrovirus genomes have been characterized by transfection into
NIH3T3 cells. The transforming activity of each of the viral DNAs
can be abolished by digestion with different site specific
restriction endonucleases in a unique manner. The reasoning
behind this experimental approach is that DNA sequences from
different cell lines could be identical, if their transforming
activity - determined by transfection - is abolished by treatment
with restriction enzymes whose recognition sites occured relatively
infrequent. Unfortunately, very few results have been published
so far regarding the effect of restriction endonuclease treatment
on transfection using DNA from chemically transformed cells. DNAs
of two independent primary transformants obtained by transfection
of NIH3T3 cells with DNA from 3-methylcholanthrene transformed
C3H10T1/2 cells showed an identical pattern of effects on
secondary transfection after cleavage with five different
restriction endonucleases (34). In contrast to these results,
DNA from transformed NIH3T3 cell lines obtained after transfection
with normal mouse DNA or with rat neuroblastoma DNA exhibited a
different pattern (34). A unique pattern of inactivation by
restriction endonucleases has been reported for DNA from three
independent NIH3T3 cell lines obtained after transfection with
DNA from normal chick embryo fibroblasts (30, see next paragraph).
Moreover, for DNA from human bladder carcinoma cell lines, a
different pattern of inactivation after treatment with
restriction endonucleases has been found (32). Thus, there
appears to be heterogeneity among transforming DNA sequences from
different chemically transformed cells. Furthermore, these DNA
sequences are probably different from transforming genes of
several murine retroviruses. However, both conclusions have to
be verified by more experimental evidence.

9. DNA FRAGMENTS OF NORMAL CELLS CAN INDUCE NEOPLASTIC TRANS-
    FORMATION IN PRENEOPLASTIC 3T3 MOUSE CELLS AT LOW
    EFFICIENCY

    During recent years, it became clear that highly oncogenic
retroviruses have integrated in their genome a cellular gene and
may transform appropriate cells by overproducing the product of
this cellular gene. For example, cells transformed with avian
sarcoma virus contain an amount of $p60^{sarc}$ protein which is about
fiftyfold higher than the amount of $p60^{src}$ protein present in
uninfected cells (35). From this observation, it was concluded
(30) that it should be possible to use DNA from normal cells for

transfection of NIH3T3 cells provided the appropriate DNA sequences which correspond to transforming genes are overexpressed after transfection. Such an overexpression could occur, if the putative transforming genes were reintegrated after transfection adjacent to an efficient promotor region of transcription. This hypothesis was supported by showing that DNA from normal chick embryo fibroblasts or from non-transformed 3T3 cells caused transformation of NIH3T3 cells, if the DNA was sheared to fragments of 0.3 to 3 x 10^6 Daltons (30). The low number of transformed clones obtained was consistent with the notion that the putative transforming gene in the normal DNA could have lost, due to the fragmentation, its flanking sequences, which caused its low expression in normal cells. Secondary transformants were obtained at high frequency using high molecular weight DNA (> 2 x 10^7 Daltons) from primary transformants. Under these conditions, the transforming gene might have been transfected together with its newly gained highly efficient promotor region.

10. LEUCOSIS VIRUS INFECTION CAN ACTIVATE CELLULAR TRANSFORMING GENES BY PROMOTOR INSERTION

Very recently, it was reported (36) that avian leucosis virus (ALV) can induce lymphomas by activating the c-myc gene, the cellular counterpart of the transforming gene of MC29 virus. As a rare event, the ALV provirus is integrated in the cellular DNA adjacent to the c-myc gene. From analysis of DNA and RNA of ALV-induced lymphomas, the authors concluded that a viral promotor caused transcription and enhanced expression of a cellular transforming gene. From these results, the possibility is suggested that activation of cellular transforming genes may be the common initiation event of neoplastic transformation by viral and non-viral agents. This suggestion is consistent with the finding that fragmented DNA of normal cells can cause neoplastic transformation after transfection. Either a transforming gene on the fragmented DNA is more efficiently transcribed after integration close to a strong promotor in the 3T3 genome, or a strong promotor region of the fragmented DNA can activate a silent transforming gene in the 3T3 genome.

11. CONCLUSIONS AND SPECULATIONS

There are two main conclusions to be drawn from studies of neoplastic transformation in somatic cell hybrids:

(i) Suppression of malignancy as well as dominant expression can be demonstrated in cell hybrids of certain tumorigenic parental cells with normal cells.

(ii) Suppression of malignancy in certain cell hybrids is likely to depend on the dosage of genes (chromosomes) in the tumorigenic versus normal parental genome.

Both conclusions from studies with somatic cell hybrids fit
into the still fragmentary mechanistic model of neoplastic transformation which emerges from studies of transfection with DNA
from tumorigenic cells. At least two different kinds of transforming lesions in the genome of tumorigenic cells can be
distinguished (30):

(i) Tumorigenic cell lines, whose DNA can be successfully used
for transfection of the transformed phenotype, could harbor a
mutation in a cis-acting regulatory sequence, for example a
promotor-up mutation. This mutated sequence may act in a dominant
fashion after transfer into 3T3 cells. No attempts or results
of transfecting normal diploid fibroblasts instead of 3T3 cells
with DNA from these tumorigenic cells have been published. Thus,
future experiments have to show whether dominant expression of
malignancy in somatic cell hybrids is caused by a similar
mechanism as suggested for dominant expression of neoplastic
transformation after transfection of 3T3 cells.

(ii) The suggestion (30) that mutational inactivation of a trans-
-regulatory diffusible molecule might be the molecular mechanism
of transformation in many tumorigenic cell lines could also
explain suppression of malignancy in somatic cell hybrids. In
such hybrids, the simultaneous production of functional trans-
-regulatory molecules (by the normal parental genome) and of
inactivated trans-regulatory molecules (by the tumorigenic parental genome) would lead to suppression of neoplastic transformation.
If the gene coding for the functional trans-regulatory molecule is
lost from these hybrids due to segregation of "normal" chromosomes,
malignancy would be reexpressed.

Alternatively, one could explain suppression of malignancy
and the dosage phenomenon in somatic cell hybrids in the following way: Neoplastic transformation in the tumorigenic parental
cell of a cell hybrid may be caused by the slightly excessive
overproduction of a normal gene product which has pleiotropic
effects, for example due to phosphorylation of several different
cellular proteins. In somatic cell hybrids of this tumorigenic
cell line with normal cells, there should be the double amount
of these latter cellular proteins present. Under these conditions, the overproduction of the critical gene product may be
insufficient to cause neoplastic transformation. Such a mechanism
could also explain that, in somatic cell hybrids, several normal
chromosomes may contribute to suppression of malignancy.
Alternatively, other mechanisms appear possible.

One can expect that further investigations on the transfer of
transforming genes or regulatory DNA sequences should provide many
new insights into the puzzling mechanism of neoplastic transformation.

12. ACKNOWLEDGEMENT

Our work on genetic control of neoplastic transformation is supported by the Deutsche Forschungsgemeinschaft (SFB 102).

13. REFERENCES

1) Ringertz, R.N. and Savage, R.E. (1976). Gene mapping and gene complementation analysis. In: Cell hybrids. Academic Press, New York, p. 224.

2) Barski, G., Sorieul, S., and Cornefert, F. (1961). "Hybrid" type cells in combined cultures of two different mammalian cell strains. J. Natl. Canc. Inst., 26, 1269.

3) Ozer, H.L. and Jha, K.K. (1977). Malignancy and transformation. Expression in somatic cell hybrids and variants. Adv. Canc. Res., 25, 53.

4) Barski, G. and Belehradek, jr.J. (1979). Inheritance of malignancy in somatic cell hybrids. Somatic Cell Genet., 5, 897.

5) Harris, H. (1979). Some thoughts about genetics, differentiation, and malignancy. Somatic Cell Genet., 5, 923.

6) Sidebottom, E. (1980). The analysis of malignancy by cell fusion. In vitro (Rockville); 16, 77.

7) Croce, C.M. (1980). Cancer genes in cell hybrids. Biochim. Biophys. Acta, 605, 411.

8) Smets, L.A. (1980). Cell transformation as a model for tumor induction and neoplastic growth. Biochim. Biophys. Acta, 605, 93.

9) Harris, H. (1979). Some recent progress in the analysis of malignancy by cell fusion. In: The Ciba Foundation Symposion No. 66, 1978. Human Genetics: Possibilities and Realities (eds., R. Porter and M. O'Connor), Excerpta Medica, Amsterdam, p. 311.

10) Bishr Omary, M., Townbridge, I.S., and Minozada, J. (1980). Human cell-surface glycoprotein with unusual properties. Nature, 286, 888.

11) Kucherlapati, R. and Tepper, R. (1978). Modulation and mapping of human plasminogen activator by cell fusion. Cell, 15, 1331.

12) Eun, C.K. and Klinger, H.P. (1980). Human chromosome 11 affects the expression of fibronectin fibers in human x mouse cell hybrids. Cytogenet. Cell Genet., 27, 57.

13) Marshall, C.J. and Dave, H. (1978). Suppression of the transformed phenotype in somatic cell hybrids. J. Cell Science, 33, 171.

14) Schäfer, R., Doehmer, J., Drüge, P.M., Rademacher, I., and Willecke, K. (1981). Genetic analysis of transformed and malignant phenotypes in somatic cell hybrids between tumorigenic Chinese hamster cells and diploid mouse fibroblasts. Cancer Res., 41, 1214.

15) Stanbridge, E.J. and Wilkinson, J. (1978). Analysis of malignancy in human cells: malignant and transformed phenotypes are under separate genetic control. Proc. Natl. Acad. Sci. U.S.A., 75, 1466.

16) Klinger, H.P. (1980). Suppression of tumorigenicity in somatic cell hybrids. I. Suppression and reexpression of tumorigenicity in diploid human x D98AH2 hybrids and independent segregation of tumorigenicity from other cell phenotypes. Cytogenet. Cell Genet., 27, 254.

17) Muggleton-Harris, A.L. and Palumbo, M. (1979). Nucleocytoplasmic interactions in experimental binucleates formed from normal and transformed components. Somatic Cell Genet., 5, 397.

18) Bunn, C.L. and Tarrant, G.M. (1980). Limited lifespan in somatic cell hybrids and cybrids. Exp. Cell Res., 127, 385.

19) Jonasson, J., Povey, S., and Harris, H. (1977). The analysis of malignancy by cell fusion. VII. Cytogenetic analysis of hybrids between malignant and diploid cells and of tumours derived from them. J. Cell Science, 24, 217.

20) Aviles, D., Jami, J., Rousset, J.-P., and Ritz, E. (1977). Tumor x host cell hybrids in the mouse: Chromosomes from the normal cell parent maintained in malignant hybrid tumors. J. Natl. Canc. Inst., 58, 1391.

21) Aviles, D., Ritz, E., and Jami, J. (1980). Chromosomes in tumors derived from mouse x diploid cell hybrids obtained in vitro. Somatic Cell Genet., 6, 171.

22) Kucherlapati, R. and Shin, S. (1979). Genetic control of tumorigenicity in interspecific mammalian cell hybrids. Cell, 16, 639.

23) Jonasson, J. and Harris, H. (1977). The analysis of malignancy by cell fusion. VIII. Evidence for the intervention of an extra-chromosomal element. J. Cell Science, 24, 255.

24) Howell, A.N. and Sager, R. (1978). Tumorigenicity and its suppression in cybrids of mouse and Chinese hamster cell lines. Proc. Natl. Acad. Sci. U.S.A., 75, 2358.

25) Stiles, C.D., Chuman, L.M., and Saier, M.H. jr. (1976). Enhancement of tumorigenicity in athymic nude mice by coinjection of tumor cells and embryonic fibroblasts. J. Cell Biol., 70, 169a.

26) Yamamoto, T., Rabinowitz, Z., and Sachs, L. (1973). Identification of the chromosomes that control malignancy. Nature New Biol., 243, 247.

27) Shih, C., Shilo, B.-Z., Goldfarb, M., Dannenberg, A., and Weinberg, R.A. (1979). Passage of phenotypes of chemically transformed cells via transfection of DNA and chromatin. Proc. Natl. Acad. Sci. U.S.A., 76, 5714.

28) Hill, M. and Hillowa, J. (1972). Virus recovery in chicken
     cells tested with Rous sarcoma cell DNA. Nature New
     Biol., 237, 35.
29) Graham, F.L. (1977). Biological activity of tumor virus
     DNA. Adv. Canc. Res., 25, 1.
30) Cooper, G.M., Okenquist, S., and Silverman, L. (1980).
     Transforming activity of DNA of chemically transformed
     and normal cells. Nature, 284, 418.
31) Shih, C., Padhy, L.C., Murray, M., and Weinberg, R.A. (1981).
     Transforming genes of carcinomas and neuroblastomas intro-
     duced into mouse fibroblasts. Nature, 290, 261.
32) Krontiris, T.G. and Cooper, G.M. (1981). Transforming
     activity of human tumor DNAs. Proc. Natl. Acad. Sci.
     U.S.A., 78, 1181.
33) Jelinek, W.R., Toomey, T.P., Leinwand, L., Duncan, C.H.,
     Biro, P.A., Choudary, P.V., Weissman, S.M., Rubin, C.M.,
     Houck, C.M., Deininger, P.L., and Schmid, C.W. (1980).
     Ubiquitous, interspersed repeated sequences in mammalian
     genomes. Proc. Natl. Acad. Sci. U.S.A., 77, 1398.
34) Shilo, B.Z. and Weinberg, R.A. (1981). Unique transforming
     gene in carcinogen-transformed mouse cells. Nature,
     289, 607.
35) Collett, M.S., Brugge, J.S., and Erickson, R.L. (1978).
     Characterization of a normal avian cell protein related
     to the Avian Sarcoma Virus transforming gene product.
     Cell, 15, 1363.
36) Hayward, W.S., Neel, B.G., and Astrin, S.M. (1981). Acti-
     vation of a cellular onc gene by promotor insertion in
     ALV-induced lymphoid leucosis. Nature, 290, 475.

ANALYSIS OF CELL SURFACE ANTIGENS WITH MONOCLONAL ANTIBODIES

Alan F. Williams

MRC Cellular Immunology Unit
Sir William Dunn School of Pathology
University of Oxford
South Parks Road, Oxford OX1 3RE, U.K.

INTRODUCTION

This article will deal with the use of monoclonal antibodies as the primary tool for the identification and analysis of complex mixtures of molecules. The analysis of the cell surface of lymphocytes will be mainly discussed but the approach outlined is equally useful for other cell types and molecular mixtures from other parts of the cell. Antigens of interest for genetic studies will be briefly discussed.

The predominant types of molecules found at cell surfaces are glycoproteins and glycolipids and these molecules are inserted into the lipid bilayer via a hydrophilic 'tail' or surface of protein or lipid. The hydrophobicity of these molecules makes them very difficult to fractionate since for their solubilization lipids must be replaced by detergent micelles and these have a major effect on the properties of the solubilized molecules.[1] Many of the conventional techniques of biochemical fractionation cannot be usefully applied in the presence of detergents and the situation is made worse because very small amounts of material are available. Another major problem is that usually there is no biological activity for membrane molecules, which can be followed in a purification procedure with solubilized molecules. The problems of identification and purification can be solved if antibodies specific to membrane molecules can be produced. A specific antibody can be used to assay a membrane molecule in the presence of detergents and also can be used to purify the molecule by affinity chromatography. Before the advent of monoclonal antibodies one could argue that a serological approach to the cell surface had

advantages.[2] With monoclonal antibodies[3] these advantages have
become overwhelming compared with any other approach.

ANTIGENS AND ANTIGENIC DETERMINANTS

   Molecules which stimulate an immune response are called antigens and in this article an antigen will be defined as a molecule
which is recognised by the binding of antibody. Only small parts
of an antigen interact with antibody. These are called antigenic
determinants and are usually protein or carbohydrate in nature.
An antigen can only stimulate an immune response if it is recognised
by circulating lymphocytes. For this to occur, parts of the antigen's structure must be different from the structures shown by
biological material in the immunized animal, since an animal's
lymphoid system is usually tolerant to self. Thus the immune
system recognises the differences between molecules and ignores
the similarities (see Ref. 4 for basic concepts). The origin of
these differences can be genetic polymorphism within a species
(allelic variants), or genetic differences between species which
have accumulated in evolution. An antigenic determinant which is
allelic within a species is called an allotypic (or polymorphic)
determinant, while a determinant which is constantly expressed on
all products of the antigenic locus within a species is called an
isotypic (or monomorphic) determinant. Antibodies to allotypic
determinants can be produced by immunization within (most commonly)
or between species, while antibodies against isotypic determinants
are usually only produced by immunization between species.

   The structural basis for protein antigenic determinants can be
illustrated by considering the Thy-1 antigens of mouse and rat
(reviewed in Ref. 5). In the mouse there are two allelic forms
called Thy-1.1 and Thy-1.2 which can be recognised by antibodies
raised by immunizing between appropriate mouse strains. The rat
has an homologous Thy-1 molecule which shares the mouse Thy-1.1
determinant, but also differs from the mouse molecule such that
antibodies to isotypic determinants of mouse or rat Thy-1 can be
raised by immunizing between the species. Parts of the amino acid
sequences of mouse and rat Thy-1 are shown below:

```
                     51               67        86    89
     Rat Thy-1       V ——— NLFSDRF ——— MC —— R ——
     Mouse Thy-1.1   I ——— TLSNQPY ——— FC —— R ——
     Mouse Thy-1.2   I ——— TLSNQPY ——— FC —— Q ——
```

(Total sequence is 112 residues in mouse and 111 in rat.
The lines indicate identities in sequence between the
species. Sequences from Ref. 6 and Williams and Gagnon,
unpublished.)

Between mouse Thy-1.1 and Thy-1.2 there is only one amino acid residue difference (arginine/glutamine at 89) which is presumably responsible for the allotypic determinant. Between rat Thy-1 and mouse Thy-1.1 there are 20 out of 112 positions which differ in sequence (some of which are shown above) and these must account for the isotypic determinants. It is interesting to note that the Thy-1.1 determinant is allotypic in the mouse but isotypic in the rat since so far no rat has been found which lacks the Thy-1.1 determinant.

For genetic studies it is important to know the chemical nature of an antigen and its antigenic determinants. If the determinants are protein then a product of a genetic locus is being directly assayed, while if they are carbohydrate the result of an enzymatic activity of a glycosyl transferase is being followed. The main antigens of plasma membranes are glycoproteins and glycolipids and for glycoproteins the antigenic determinants could be protein or carbohydrate. However the carbohydrate of mammalian glycoprotein seems poorly antigenic and there are no well established cases for the antigenic determinants of typical membrane glycoproteins being carbohydrate. Unfortunately it is very difficult to establish unequivocally that a determinant is protein-based without detailed structural analysis. In simple tests protein determinants should be destroyed by proteolysis and in some cases by denaturation. Carbohydrate determinants should resist these treatments and be destroyed by removal of carbohydrate (very difficult to do) and be lost when a glycoprotein is biosynthesized in the presence of tunicamycin (if N-linked structures are involved).

MONOCLONAL ANTIBODIES

Most membrane glycoproteins will be antigens if another species is immunized since most polypeptide sequences have diverged in evolution to provide a basis for isotypic antigenic determinants (see above). Thus potentially an antibody can be made to any membrane glycoprotein. However, as shown in Fig. 1, with conventional techniques specific antibodies can only be produced by immunization within a species to identify allotypic determinants[7], or by immunization with pure antigen between species. In the first case specificity is controlled by genetic manipulation (isolation of a determinant by backcrossing to give congenic strains) and in the second by biochemical criteria. However neither method solves the problem of a ready supply of specific antibodies for the analysis of complex mixtures of molecules.

The specificity problem is solved with monoclonal antibodies[8,9] since in this case the primary source of material is a single antibody-secreting cell in the spleen and not a mixture of anti-

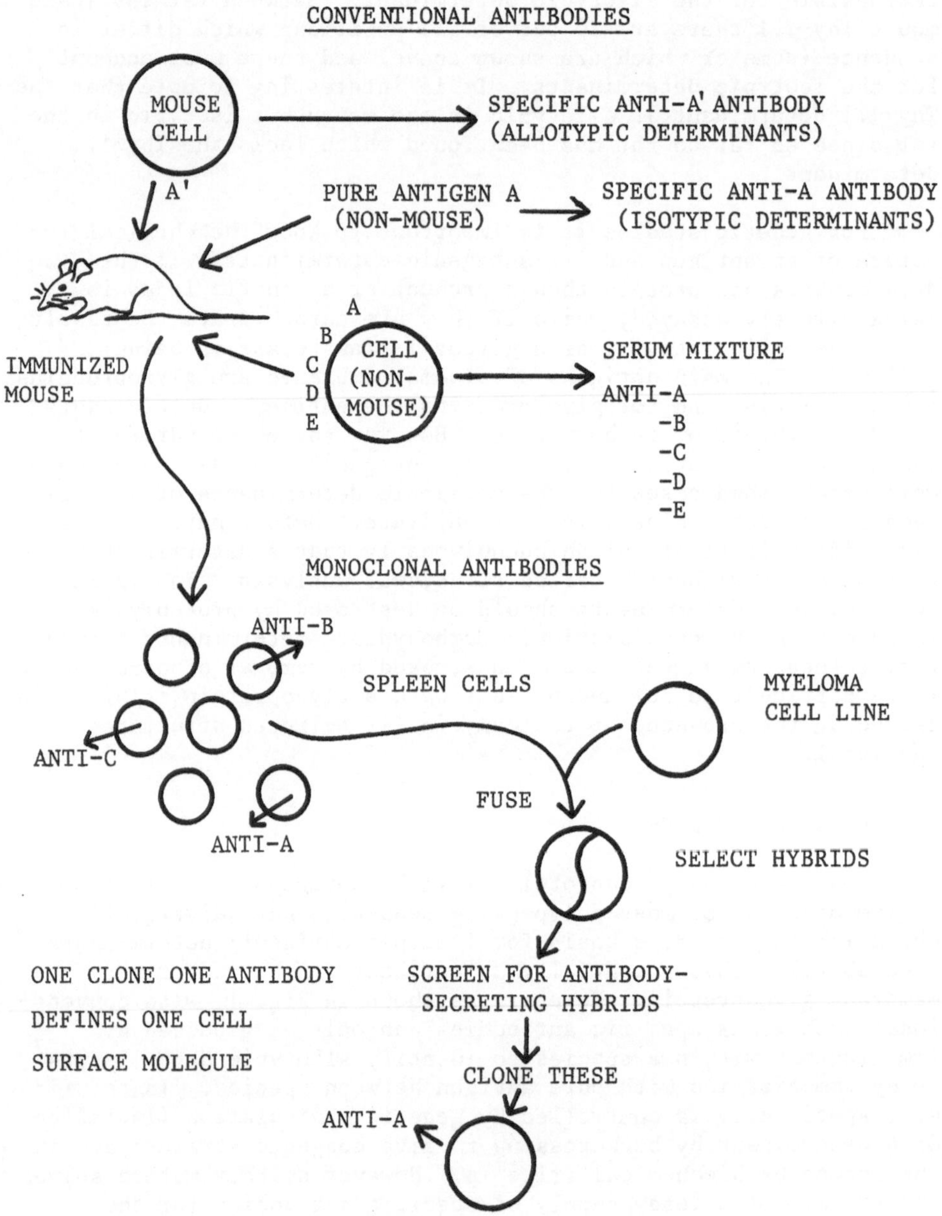

CONVENTIONAL ANTIBODIES

MOUSE CELL → SPECIFIC ANTI-A' ANTIBODY (ALLOTYPIC DETERMINANTS)

A'

PURE ANTIGEN A (NON-MOUSE) → SPECIFIC ANTI-A ANTIBODY (ISOTYPIC DETERMINANTS)

IMMUNIZED MOUSE

A B C D E CELL (NON-MOUSE) → SERUM MIXTURE

ANTI-A
    -B
    -C
    -D
    -E

MONOCLONAL ANTIBODIES

ANTI-B

SPLEEN CELLS

MYELOMA CELL LINE

ANTI-C

FUSE

ANTI-A

SELECT HYBRIDS

ONE CLONE ONE ANTIBODY DEFINES ONE CELL SURFACE MOLECULE

SCREEN FOR ANTIBODY-SECRETING HYBRIDS

CLONE THESE

ANTI-A

Fig. 1 Conventional immunization compared with production of monoclonal antibodies (Refs. 2,3,9)

body molecules in the serum (Fig. 1). Spleen cells from a mouse immunized with a mixture of molecules from another species are fused with a mouse myeloma cell line and hybrid cells are selected for. These are screened for antibody production and those making antibodies with interesting specificity are cloned. When a cloned cell line is obtained, specificity is assured, because the hybrid is ultimately derived from a single antibody-forming cell which makes one antibody. The antibody can then be used to identify a molecule whose existence may have been previously unknown.

The methods for production of antibody-secreting hybrids have been reviewed in detail elsewhere (Refs. 3,10) but in brief the following points are relevant. 1. The cell line used for hybrid formation must be from a myeloma if antibody secretion is to be maintained, and must have a genetic defect so that its growth can be prevented and hybrids selected for because the defect is complemented by the spleen cell genome. 2. Fusions between lymphoid cells and a myeloma cell line of the same species are usually required. Otherwise the hybrids shed chromosomes at a rate such that stable lines cannot be maintained. Myeloma cell lines of mouse, rat and human origin are now available and monoclonal antibodies can be prepared from these species. One exception to the like-with-like rule is that stable lines can be prepared from fusions between rat and mouse cells.[11] 3. It is not essential that the myeloma cell line used in the fusion secretes antibody, on the contrary it is an advantage if it does not since then the only immunoglobulin produced is that of the antibody-secreting cell.[12] 4. The strongest possible immunization is needed since the chances of obtaining a desired antibody-secreting hybrid will depend on the frequency of the appropriate antibody-secreting cell in the spleen cell mixture. 5. Cloned hybrid cell lines can be grown indefinitely in culture (subject to recloning) and can be stored frozen in liquid nitrogen.

Apart from solving the specificity problem monoclonal antibodies have the advantage over conventional serology that large amounts of antibody can be prepared. A cloned cell line grown in tissue culture will yield up to 100µg of antibody per ml of medium and if hybrids are grown as ascitic tumours in mice about 25mgm of antibody can be obtained per mouse (5ml of ascites fluid at 5mg/ml). Thus grams of an antibody can be easily prepared and this is important in a variety of applications, for example when antibodies are to be used for affinity chromatography.

SEROLOGICAL ASSAYS

In making and using monoclonal antibodies it is essential that suitable assays are used. In particular screening methods which quickly identify desirable antibodies are needed if time is not to

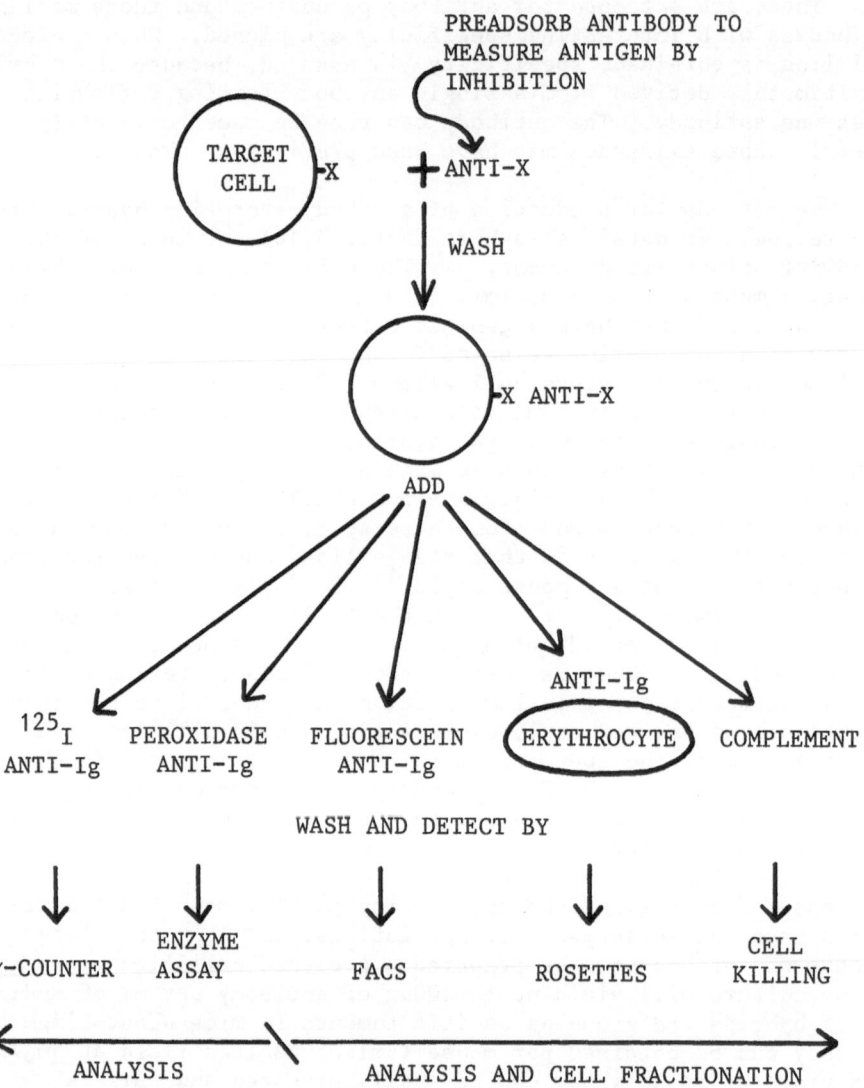

Fig. 2 Assays for analysis of cell surface antigens

References: ^{125}I-labelled anti-Ig[2], peroxidase anti-Ig[14], fluorescein anti-Ig[15], rosette assays[16], cytotoxicity[17]

be wasted in cloning hybrids which will ultimately not be useful. Assaying monoclonal antibodies can be quite different from assaying conventional antibodies since usually a monoclonal antibody will react with only one determinant on an antigen. In contrast to this a conventional antiserum specific for one antigen will contain a mixture of antibodies reacting with different determinants on the antigen. This is favourable for traditional assays like the precipitin reaction or complement fixation or killing. These assays depend on cross-linking of antigens, or the localization of a number of antibody molecules in one area, and generally are not the best ones for assaying monoclonal antibodies.

In all serological assays ultimately the aim is to measure the binding of antibody to antigen and this can be simply done as shown in Fig. 2. All the assays shown, with the exception of complement-mediated killing, are indirect binding assays where the binding of first antibody is indicated by interaction with purified anti-immunoglobulin antibody tagged with various marker materials. Similar assays are possible with soluble antigens if these are absorbed to plastic plates the surface of which becomes equivalent to the particulate antigen of a cell surface. The assay of choice will depend on the aim of the experiment and in particular will differ depending on whether or not antigen on individual cells is to be measured (discussed in detail in Refs. 2,13).

If the aim is to measure total antigen throughout a purification or in various biological preparations then a quantitative radioactive or enzyme-linked binding assay should be used and the presence of antigen measured by preincubating the first antibody with test material and assaying the inhibition of antibody binding. Assays can be done in the presence of detergents if gluteraldehyde-fixed cells are used. For assaying antigen on individual cells the enzyme-linked method is most favourable if tissue sections[18] are being studied, and the fluorescence method is best for single cells if analysis with a fluorescence-activated cell sorter[15] (FACS) is possible.

The assays involving antibody tagged with ^{125}I, enzymes or fluorescein differ from rosetting and cytotoxicity assays in that in the former three antibody binding can be quantitated, while in the latter two the assay is non-quantitative. The non-quantitative assays register nothing below a threshold of binding and above this a reaction is scored. The effect is all or none and this presents such disadvantages for analysis that there is no obvious reason why these tests should be used in most analytical work.

BINDING CHARACTERISTICS OF ANTIBODIES

Serology is complicated by the fact that antibodies are multi-

valent (bivalent, IgG; tetravalent, IgA; decavalent, IgM) and antigens may be in a univalent or multivalent form. If an antibody
reacts with a multivalent antigen then the dissociation rate will be
greatly reduced compared with a univalent reaction[19,20] and this has
important consequences in serology. In many cases an inhibiton assay
may be done using inhibiting antigen in univalent form with the assay
for residual antibody carried out on a multivalent antigenic particle
(cell) or surface (plate assay) (Fig. 2). If the reaction were to
come to equilibrium then inhibition by univalent antigen would not
be seen since the multivalent target would acquire most of the antibody due to the reduced dissociation rate of a bivalently bound antibody (Fig. 3). If the dissociation rate for the univalent interaction
is fast compared with the time of the assay then the form of the antigen greatly affects the outcome of the serological test (Fig. 3).
Thus the kinetics of interaction are all-important.

Studies on hapten-antibody interactions suggested that the kinetics of antibody binding were extremely rapid.[19] However, with real
antigens the on and off rates are much slower and for example a mouse
anti-Thy-1.1 antibody with an equilibrium constant of 4×10^9 M^{-1}
for the univalent interaction had a $t_{\frac{1}{2}}$ for univalent dissociation of
about 150 min.[20] With such an antibody the inhibition is not greatly
affected by the form of the antigen since the incubation time with
target cells in the binding assay is 30 to 60 min. Fortunately antibodies like this are commonly produced after hyperimmunization and
it can be argued that they should be sought in primary screening.
This can be fairly easily done for membrane antigens by seeing if
solubilization in detergent causes an apparent loss of inhibitory
antigenic activity. If not then the antibody involved probably has
a slow univalent dissociation rate.

CELL SEPARATION ON THE BASIS OF ANTIBODY BINDING

Three of the serological methods in Fig. 2 can be used as a
basis of cell separation and to these can be added the binding of
cells to beads or surfaces which have been coated with antibody. The
characteristics of the methods are as follows.

Cytotoxicity[17]

In this method cells which bind antibody are lysed with complement. Thus only antigen-negative cells can be recovered. Large
numbers of cells can be handled but a disadvantage is that in many
cases antigen+ cells will not be killed because the threshold conditions for cell lysis are not met. Cytotoxicity has been successfully used in selecting for mutant cell lines which have lost
antigenic expression.[21,22]

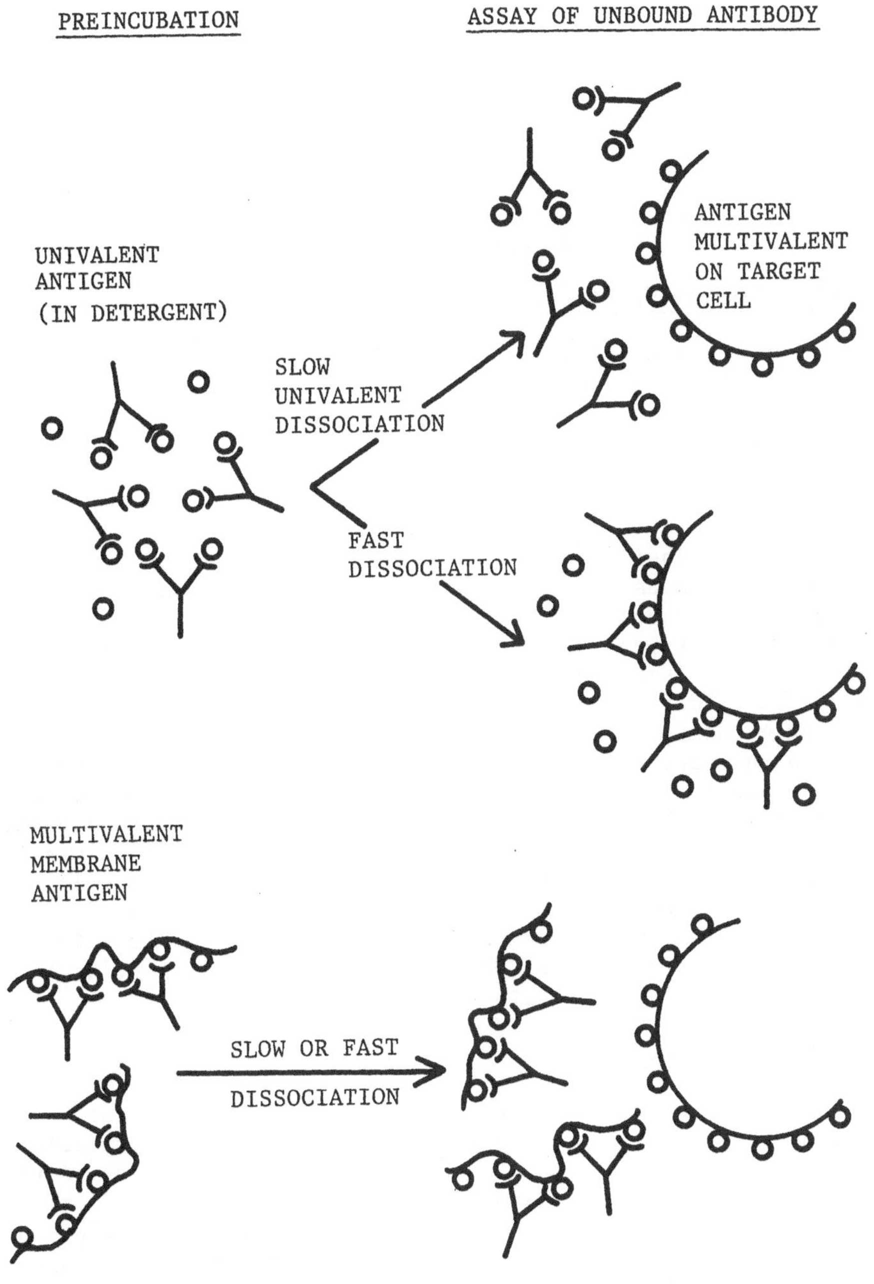

Fig. 3 Effect of dissociation rate on the inhibition of antibody binding with univalent or multivalent antigen

Rosetting[16]

In this method cells form rosettes with erythrocytes coated with antibody and can then be separated from unrosetted cells because the rosettes are larger or more dense than the single cells. In principle the rosetted and unrosetted cells could be recovered but in practice the antigen[+] population is usually contaminated with antigen[−] cells. Large cell numbers can be used and only small amounts of antigen are needed for rosette formation. Thus this method is a good one for depletion of antigen[+] cells.

Affinity Columns or Plates[23,24]

Antigen[+] cells are bound to antibody-coated beads or plastic dishes. Both bound and unbound fractions of cells can be recovered and the plastic dish method[24] may be a good one for enriching rare antigen[+] cells.

The Fluorescence-Activated Cell Sorter[15]

Cells are labelled with fluorescein-conjugated antibody and flow in a stream past a laser beam. Fluorescence emission is quantitated and cells can be sorted on the basis of the amount of fluorescence by electrical deflection of a droplet which contains the cell and which is given a positive or negative charge. The operator chooses the fluorescence emission level which activates the + or − charging mechanism and thus this is the only fractionation method in which the separation can be based on the amount of antibody bound. In all other methods the cut-off for designation of antigen[+] or [−] cells is inherent in the technique. With the FACS both antigen[+] or antigen[−] cells are recovered and the antigen[−] fraction has less contamination than the antigen[+] one (a droplet with two cells can register as antigen[+] if only one cell is antigen[+]). The disadvantages of the FACS are expense and the fact that only about 10^7 cells/hour can be sorted. The FACS has been used to select cell lines expressing antigenic variants.[25]

PURIFICATION OF ANTIGENS[26]

In Fig. 1 it is shown that by preparing monoclonal antibodies the identification of previously unknown molecules is possible. The antibody also provides a means of purifying the antigen. Antibodies are coupled to Sepharose 4B by the CNBr method and membrane molecules solubilized in non-ionic detergents or deoxycholate are passed through the column. The column is then washed in buffer with detergent and eluted with a dissociating buffer. High pH seems to be the best method of dissociation and can be used with non-ionic detergents or deoxycholate. The eluting medium is neutralized immediately after the fractions emerge from the column.

The results of purification with antibody columns have been surprisingly good and the method has worked with recoveries of at least 50% for four rat lymphoid antigens, notably LC, Ia, Thy-1 and W3/13 antigens (reviewed in Refs. 26,27). For the purification of W3/13 antigen 500µg of glycoprotein was purified from about 3×10^{11} thymocytes with an overall purification of 7000-fold. In this case the antibody column affinity step was carried out twice and this antigen could not have been purified if the antibody affinity column had not been used.[27]

SCHEME FOR ANALYSIS OF STRUCTURE AND FUNCTION OF MEMBRANE MOLECULES

Fig. 4 summarizes the monoclonal antibody approach. The anti-bodies can be used to identify and purify previously unkown molecules. Antibodies to polymorphic determinants and antigens which mark lymphocyte sub-sets can be screened for. The functions of the cell surface molecules can be approached in two ways. Firstly the effect of antibodies or pure antigens on functional systems can be assayed. It is clear that the simple attachment of an antibody to a cell surface does not routinely have functional effects. Thus the observation of inhibition[28] or stimulation[29] of cell functions by antibody suggests that the antigens involved may have important functional roles. Secondly if cell lines which show immune function-al activities are available then selection of antigen minus cells may allow identification of functionally important antigens. This could be achieved by correlating the loss of function with the loss of a particular cell surface molecule.[30]

LYMPHOID CELL SURFACE ANTIGENS SUITED TO GENETIC ANALYSIS

There are now a number of lymphoid surface antigens which are well enough characterized to be favourable subjects for genetic analysis. The advantages of studying lymphoid antigens are that cell lines which retain the differentiated properties of B and T cells are available and the products being studied can be assayed at a single cell level. Mutations resulting in loss of or altered antigen expression can be selected for by immunological techniques. To select for cells which have lost antigen expression antigen[+] cells are eliminated by complement-mediated cytotoxicity[21] or cell sorting. Antigenically altered variants can be selected by using two mono-clonal antibodies labelled with rhodamine or fluorescein which label different determinants on a molecule and sorting for cells which label for one fluorescent dye only.[25] Alternatively a change in structure can be detected with monoclonal antibodies specific for different determinants on the same molecule which sterically inhibit each others binding. If one antibody is fluorescein-labelled and incubated with cells in an excess of the other unlabelled antibody then the fluorescein antibody will only bind to cells with a change in the antigen such that the unlabelled antibody will not bind.[25]

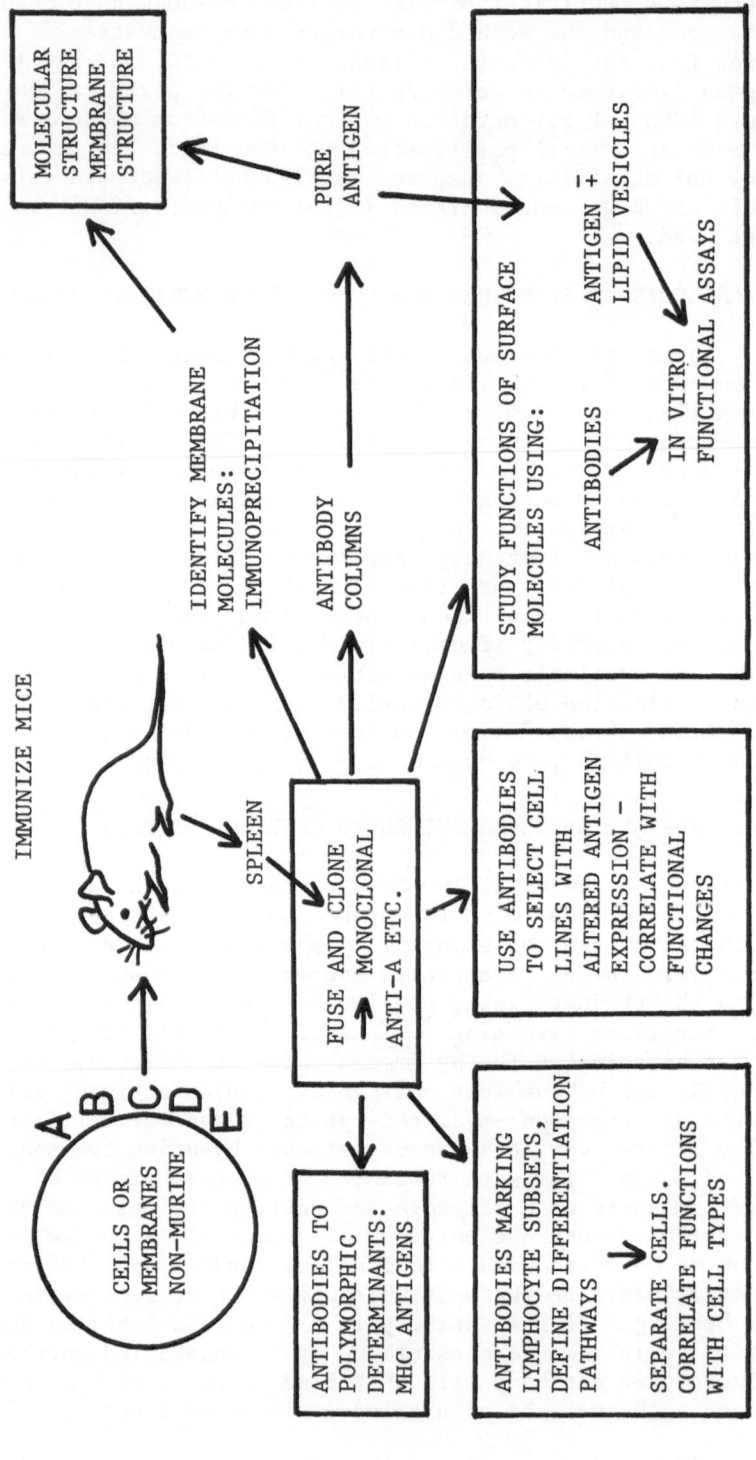

Fig. 4 Using monoclonal antibodies to analyse structure and function of cell surface molecules

The labelled cells can then be enriched with the FACS.

The following antigens can be studied:

Immunoglobulin

Mutants have been sought in an attempt to understand the structure heterogeneity of immunoglobulin.[31] Immunoglobulin has usually been studied as a secreted product and not as a cell surface antigen.

Class I and Class II Histocompatibility Antigens

Class I antigens consist of a non-covalently associated dimer of polypeptides of 43,000 and 12,000 mol. wt. Human HLA[32] and mouse H-2[33] antigens have now been sequenced. Structural variants are of interest since these antigens show great allelic variation in the large polypeptide within a species and both polypeptide chains show sequence homology with immunoglobulin.[32] Mutants of H-2 antigens arising in vivo have been studied[34] and structural variants in cell lines can be selected for as described above.[25]

The Class II or Ia histocompatibility antigens consist of polypeptides of 28,000 and 32,000 mol. wt. which are not yet sequenced.[35] The 28,000 mol. wt. polypeptide shows extensive polymorphism and structural variants of this selected in cell lines would be of interest. Also the Class II antigens would be suitable for studying as a differentiated cell product since they are found only on B lymphocytes, dendritic cells, activated macrophages and certain epithelial cells.

Many monoclonal antibodies are available against histocompatibility antigens of mice, rats and humans.[36]

Thy-1 Antigens

Thy-1 is a major glycoprotein of thymocytes and brain in rodents and is a polypeptide of 12,500 mol. wt. with three carbohydrate structures.[6] Thy-1 is homologous to an immunoglobulin domain and may be like the primordial Ig domain.[37] Structural variants in Thy-1 would thus be of interest. Thy-1 minus mutants have been extensively studied by Hyman and Trowbridge and a number of these involve glycosylation defects.[38] The Thy-1 antigen may be a very favourable molecule to study with respect to the regulation of the differentiated state. It has been shown that Thy-1 expression is extinguished if T cell lines are fused with B cell lines.[39] The extent of this effect can vary with the cell line and also with the use of pseudodiploid or pseudotetraploid cell lines as fusing partners to the T cell line.[40]

TL Antigen, T-200 (Leucocyte Common) Antigen, and Ly-2,3 Antigen

TL, T-200 and Ly-2,3 antigens are all differentiation antigens which are less well characterized than the above antigens. Antigen minus cell lines have been selected for these antigens (TL[41], T-200[42], Ly-2,3[30]). For T-200 two types of mutant lines were obtained; one is possibly defective in the structural gene for mouse T-200, the other for a gene involved in regulation or biosynthesis.[42]

ACKNOWLEDGEMENTS   I am grateful to Mrs Gaynor Newton for drawing the figures and to Mrs Christine Scott for typing the manuscript.

REFERENCES

1. A. Helenius and K. Simons, Solubilization of membranes by detergents, Biochim. Biophys. Acta 415:29 (1975).
2. A. F. Williams, Differentiation antigens of the lymphocyte cell surface, in "Contemporary Topics in Molecular Immunology 6:83, G. L. Ada and R. R. Porter, eds., Plenum Press, New York (1977).
3. C. Milstein, Monoclonal antibodies, Scientific American 243:56 (1980)
4. L. E. Hood, I. L. Weissman and W. B. Wood, "Immunology", The Benjamin/Cummings Publ. Co. Inc., 2727 Sand Hill Road, Menlo Park, California 94025 (1978).
5. A. F. Williams, A. N. Barclay, M. Letarte-Muirhead and R. J. Morris, Rat Thy-1 antigens from thymus and brain: their tissue distribution, purification and chemical composition, Cold Spring Harbor Symp. Quant. Biol. 41:51 (1976).
6. D. G. Campbell, J. Gagnon, K. B. M. Reid and A. F. Williams, Rat brain Thy-1 glycoprotein. The amino acid sequence, disulphide bonds and an unusual hydrophobic region. Biochem.J. 195:15 (1981).
7. E. A. Boyse and L. J. Old, Some aspects of normal and abnormal cell surface genetics, Ann. Rev. Gent. 3:269 (1969).
8. G. Köhler and C. Milstein, Derivation of specific antibody-producing tissue culture and tumor lines by cell fusion, Eur. J. Immunol. 6:511 (1976).
9. A. F. Williams, G. Galfrè and C. Milstein, Analysis of cell surfaces by xenogeneic myeloma-hybrid antibodies: differentiation antigens of rat lymphocytes, Cell, 12:663 (1977).
10. R. H. Kennett, T. J. McKearn and K. B. Bechtol, eds., "Monoclonal Antibodies. Hybridomas: A New Dimension in Biological Analyses", Plenum Press, New York and London.(1980).
11. J. Schroder, K. Autio, J. M. Jarvis and C. Milstein, Chromosome segregation and expression of rat immunoglobulin in rat/mouse hybrid myelomas, Immunogenetics 10:125 (1980).

12. J. F. Kearney, A. Radbruch, B. Liesegang and K. Rajewsky, A new mouse myeloma cell line that has lost immunoglobulin expression but permits the construction of antibody-secreting hybrid cell lines, J. Immunol. 123:1548 (1979).

13. A. F. Williams, Cell-surface antigens of lymphocytes: markers and molecules, Biochem. Soc. Symp. 45:27 (1980).

14. P. K. Nakane, Recent progress in the peroxidase-labeled antibody method, Ann. N.Y. Acad. Sci. 254:203 (1975).

15. L. A. Herzenberg and L. A. Herzenberg, Analysis and separation using the fluorescence activated cell sorter (FACS) in "Handbook of Experimental Immunology", 3rd edn., D. M. Weir, ed., Blackwell Scientific Publications, Oxford (1978).

16. C. Parish and J. Hayward, The lymphocyte surface. I. Relation between Fc receptors, C'3 receptors and surface immunoglobulins, Proc. Roy. Soc. Lond. B 187:47 (1974).

17. J. Klein, "Biology of the Mouse Histocompatibility-2 Complex", Springer-Verlag, New York (1975).

18. A. N. Barclay, The localization of populations of lymphocytes defined by monoclonal antibodies in rat lymphoid tissues, Immunology 42:593 (1981).

19. F. Karush, The affinity of antibody: range, variability, and the role of multivalence, in "Comprehensive Immunology" 5, R. A. Good and G. W. Litman, eds., Plenum Medical, New York and London (1978).

20. D. W. Mason and A. F. Williams, The kinetics of antibody binding to membrane antigens in solution and at the cell surface, Biochem. J. 187:1 (1980).

21. R. Hyman, Studies on surface antigen variants. Isolation of two complementary variants for Thy 1.2, J. Natl. Cancer Inst. 50:415 (1973).

22. T. V. Rajan, H-2 antigen variants in a cultured heterozygous mouse leukemia cell line, Immunogenetics 10:423 (1980).

23. E. D. Crum and D. D. McGregor, Functional properties of T and B cells isolated by affinity chormatography from rat thoracic duct lymph, Cellular Immunology 23:211 (1976).

24. L. J. Wysocki and V. L. Sato, 'Panning' for lymphocytes: A method for cell selection, Proc. Natl. Acad. Sci. U.S.A. 75:2844 (1978).

25. B. Holtkamp, M. Cramer, H. Lemke and K. Rajewsky, Isolation of a cloned cell line expressing variant H-2K^k using fluorescence-activated cell sorting, Nature 289:66 (1981).

26. W. R. McMaster and A. F. Williams, Monoclonal antibodies to Ia antigens from rat thymus, Immunological Reviews 47:117 (1979).

27. W. R. A. Brown, A. N. Barclay, C. A. Sunderland and A. F. Williams, Identification of a glycophorin-like molecule at the cell surface of rat thymocytes, Nature 289:456 (1981).

28. M. Webb, D. W. Mason and A. F. Williams, Inhibition of the mixed lymphocyte response with a monoclonal antibody specific for a rat T lymphocyte subset, Nature 282:841 (1979).

29. J. P. Van Wauwe, J. R. De Mey and J. G. Goossens, OKT3: A mono-
    clonal anti-human T lymphocyte antibody with potent mitogenic
    properties, J. Immunol. 124:2708 (1980).
30. D. P. Dialynas, M. R. Loken, A. L. Glasebrook and F. W. Fitch,
    Lyt-2⁻/Lyt-3⁻ variants of a cloned cytolytic T cell line lack
    antigen receptor function in cytolysis, J. Exp. Med. 153:595
    (1981).
31. K. Adetugbo, C. Milstein and D. S. Secher, Molecular analysis of
    spontaneous somatic mutants, Nature 265:299 (1977).
32. H. T. Orr, D. Lancet, R. J. Robb, J. A. Lopez de Castro and
    J. L. Strominger, The heavy chain of human histocompatibility
    antigen HLA-B7 contains an immunoglobulin-like region, Nature
    282:266 (1979).
33. J. E. Coligan, T. J. Kindt, H. Uehara, J. Martinko and S. G.
    Nathenson, Primary structure of a murine transplantation
    antigen, Nature 291:35 (1981).
34. R. Nairn, K. Yamaga and S. G. Nathenson, Biochemistry of the
    gene products from murine MHC mutants, Ann. Rev. Genet.
    14:241 (1980).
35. J. Klein, The major histocompatibility complex of the mouse,
    Science 203:516 (1979).
36. G. Moller, ed., "Hybrid Myeloma Monoclonal Antibodies Against
    MHC Products" Immunological Reviews 47, Munksgaard,
    Copenhagen (1979).
37. F. E. Cohen, J. Novotný, M. J. E. Sternberg, D. G. Campbell and
    A. F. Williams, Analysis of structural similarities between
    brain Thy-1 antigen and immunoglobulin domains. Evidence for
    an evolutionary relationship and a hypothesis for its
    functional significance, Biochem. J. 195:31 (1981).
38. R. Hyman and I. Trowbridge, Analysis of the biosynthesis of T25
    (Thy-1) in mutant lymphoma cells, in "Differentiation of
    Normal and Neoplastic Hematopoietic Cells", B. Clarkson,
    P. Marks and J. Till, eds., Cold Spring Harbor Lab, Cold
    Spring Harbor, New York (1978).
39. R. Hyman and V. Stallings, Evidence for a gene extinguishing cell-
    surface expression of the Thy-1 antigen, Immunogenetics 6:447
    (1978).
40. R. Hyman, K. Cunningham and V. Stallings, Effect of gene dosage
    on cell-surface expression of Thy-1 antigen in somatic cell
    hybrids between Thy-1⁻ Abelson-leukemia-virus induced lymph-
    omas and Thy-1⁺ mouse lymphomas, Immunogenetics 12:381 (1981).
41. R. Hyman and V. Stallings, Characterization of a TL-variant of
    a homozygous TL⁺ mouse lymphoma, Immunogenetics 3:75 (1976).
42. R. Hyman and I Trowbridge, Two complementation classes of T200
    (Ly-5) glycoprotein-negative mutants, Immunogenetics 12:511
    (1981).

CONTRIBUTORS

BRAVO, R., Division of Biostructural Chemistry, Department of Chemistry, Aarhus University, Denmark. p. 43.

BRÁZ, J., Instituto Gulbenkian de Ciência, Oeiras, Portugal. p. 69.

BUTTIN, G., Institut de Recherche en Biologie Moléculaire (CNRS), Paris, France. p. 1.

CASKEY, C.T., Howard Hughes Medical Institute, Department of Medicine and Biochemistry, Baylor College of Medicine, Houston, Texas, U.S.A. p. 19.

CELIS, J.E., Division of Biostructural Chemistry, Department of Chemistry, Aarhus University, Denmark. p. 43.

CROCE, C.M., The Wistar Institute of Anatomy and Biology, Philadelphia, Pennsylvania, U.S.A. p. 55.

DEBATISSE, M., Institut de Recherche en Biologie Moléculaire (CNRS), Paris, France. p. 1.

DOLBY, T.W., The Wistar Institute of Anatomy and Biology, Philadelphia, Pennsylvania, U.S.A. p. 55.

FENWICK, Jr., R.G., Howard Hughes Medical Institute, Department of Medicine and Cell Biology, Baylor College of Medicine, Houston, Texas, U.S.A. p. 19.

FEY, S.J., Division of Biostructural Chemistry, Department of Chemistry, Aarhus University, Denmark. p. 43.

FREIRE, M.T., Instituto Gulbenkian de Ciência, Oeiras, Portugal. p. 69.

HABER, D.A., Department of Biological Sciences, Stanford University, Stanford, California, U.S.A. p. 97.

El KAREH, A., Columbia University, College of Physicians and Surgeons, New York, New York, U.S.A. p. 111.

KONECKI, D.S., Howard Hughes Medical Institute, Baylor College of Medicine, Houston, Texas, U.S.A. p. 19.

KOPROWSKI, H., The Wistar Institute of Anatomy and Biology, Philadelphia, Pennsylvania, U.S.A. p. 55.

LECHNER, M.C., Instituto Gulbenkian de Ciência, Oeiras, Portugal. p. 69.

LINNENBACH, A., The Wistar Institute of Anatomy and Biology, Philadelphia, Pennsylvania, U.S.A. p. 55.

MACDONALD-BRAVO, H., Division of Biostructural Chemistry, Department of Chemistry, Aarhus University, Denmark. p. 43.

OSTRANDER, M., Columbia University, College of Physicians and Sur-
  geons, New York, New York, U.S.A. p. 111.
RAY, P.N., Department of Genetics, Hospital for Sick Children,
  Toronto, Ontario, Canada. p.127.
RUDDLE, F.H., Yale University, Department of Biology, New Haven,
  Connecticut, U.S.A. p. 87.
de SAINT VINCENT, B.R., Institut de Recherche en Biologie Molécu-
  laire (CNRS), Paris, France. p. 1.
SCHÄFER, R., Institut für Zellbiologie (Tumorforschung), Universität
  Essen, Federal Republic of Germany. p. 43, p. 183.
SCHIMKE, R.T., Department of Biological Sciences, Stanford Univer-
  sity, Stanford, California, U.S.A. p. 97.
SILVERSTEIN, S., Columbia University, College of Physicians and Sur-
  geons, New York, New York, U.S.A. p. 111.
SIMINOVTICH, L., Department of Genetics, Hospital for Sick Children,
  Toronto, Ontario, Canada. p.127.
SINOGAS, C.M., Instituto Gulbenkian de Ciência, Oeiras, Portugal.
  p. 69.
WEISS, M.C., Centre de Génétique Moléculaire du C.N.R.S., Gif-Sur-
  Yvette, France. p. 169.
WILLECKE, K., Institut für Zellbiologie (Tumorforschung), Universi-
  tät, Essen, Federal Republic of Germany. p. 43, p. 183.
WILLIAMS, A.F., MRC Cellular Immunology Unit, Sir William Dunn
  School of Pathology, University of Oxford, Oxford, United
  Kingdom. p. 197.

INDEX